Proactive Condition Monitoring of Low-Speed Machines

Zhaklina Stamboliska · Eugeniusz Rusiński
Przemyslaw Moczko

Proactive Condition Monitoring of Low-Speed Machines

Springer

Zhaklina Stamboliska
TITAN Group
Usje Cement Plant
Skopje
Macedonia

Eugeniusz Rusiński
Przemyslaw Moczko
Department of Machine Design
and Research
Wroclaw University of Technology
Wroclaw
Poland

ISBN 978-3-319-10493-5 ISBN 978-3-319-10494-2 (eBook)
DOI 10.1007/978-3-319-10494-2

Library of Congress Control Number: 2014951723

Springer Cham Heidelberg New York Dordrecht London

Printed on acid-free paper

Springer is part of Springer Science+Business Media (www.springer.com)

Preface

In the last 10 years the industry has undergone major changes from the maximalist approaches in production and maintenance till 2008 when World industry was booming, down to minimalistic approach with major costs saving and reduction of all expenses related to research and development, maintenance, trainings, etc., during the last World crisis. The authors of this book have witnessed these difficult changes from inside the industrial facilities and outside from the external consultancy and research aspect. During that time, the maintenance of machinery has suffered a lot, especially condition monitoring was in most cases the first to be cut, but also the most re-calling maintenance aspect after every major breakdown.

In heavy duty industries, most of the core equipment are low-speed machines and their health is of vital importance for the overall operability of the companies. Yet still, very seldom can a company be found with properly established condition monitoring system.

About 10 years ago research was started in the area of condition monitoring of such low-speed machines in the heavy duty industry covering coal and copper mines, cement plant, and ferronickel production facilities. The authors were surprised that the level of condition monitoring in most cases was much below the level of, for example, automotive industry, power industry, etc. It was also found that for these machines, operating at low speeds as 100, 20 rpm…., the general principles of condition monitoring for high-speed machines are not valid and there has to be new theory that will cover organization of monitoring practices related to stationary and nonstationary operating conditions, specialized measuring systems appropriate for the low energy emitted by the discrepancies in the machines components up to specialized and tailor-made analyzing tools for assessment of machinery health.

One of the advantages of low-speed machines that were recognized was the realistic and easier performance of proactive thinking and approaching machines maintenance in terms of monitoring condition of factors that affect the main components of machines. For example, it was found that we could easily monitor the load for specialized mining equipment such as excavators, reclaimers, etc., and transfer "effects" in terms of impact to the bearings health, so by "allowing" proper load, bearings could obtain longer lifetime. Or, a similar case would be if we could establish monitoring of the cement kiln crank phenomena, we could predict and prevent failures as hot bearings, loss of contact at supporting rollers, etc.

Moreover, using FEM modeling, we could also create models of major machines and simulate various cases from exploitation, thus determining certain limit values of parameters that could be monitored and measured in real industrial facilities. The FEM usage has replaced many expensive tests that would usually be necessary to be performed in order to determine some important parameters. This approach is also useful in unusual cases that are not allowed to happen for testing purposes in real operating condition due to huge costs or critical risks.

With all these facts in our minds, we decided to develop a new approach that we called a *Proactive Condition Monitoring of Low-Speed Machines*. Not to confuse the readers, it is not a whole system of proactive maintenance, but "just" a proactive condition monitoring. The principles and advantages of this approach are described in the book. Furthermore, throughout this research and with the many contacts with industrial engineers and managers working in maintenance and production departments, it was seen that there is a lack of specialized literature that would summarize all the main aspects on condition monitoring of these low-speed machines. This has put in front of us an objective to provide better support for the industry and help technical personnel to create their own effective condition monitoring systems within a short time for their major machines.

Therefore, the book contains review of currently available condition monitoring techniques applicable for low-speed machines with recommendations on how to select the appropriate one for particular machines. The best practices for technical assessment of machinery health are given and special insight is made into the peculiarities of low-speed machines and their successful condition monitoring. Application of FEM in condition monitoring as a very helpful tool is described and at the end, a case study on application of this approach for condition monitoring of cement rotary kiln is given.

We are thankful to all engineers and researchers, but also to the many experienced technicians in the field, who have accompanied our condition monitoring journey and we sincerely hope they will all benefit from the content of this book.

Zhaklina Stamboliska
Eugeniusz Rusiński
Przemyslaw Moczko

Contents

Introduction—Review of Today's Industry and Role of Condition Monitoring

1

Abstract

Brief review of the condition, the new developments and trends in industry and related researches is given. Difference of today's maintenance process and system with the adequate systems only a decade ago is given appointing to the fast-growing demands and responsibilities. The main reasons for the development of separate condition monitoring systems for low-speed machinery are given. A short intro of the contents of next chapters is also given.

Keywords

Condition monitoring · Maintenance · Proactive approach · Slow changes · Equipment condition · FMCEA · FEM

Today, modern maintenance is in a phase of change from a pure maintenance prevention-prediction to a pro-active and operators-driven maintenance concept. It is simply not enough to prevent and predict failures, whether potential or functional. The aim is to build systems where maintenance and operators will join their know-how and actions to provide a proactive condition monitoring system where even the smallest correction will be undertaken on time to eliminate factors negatively affecting equipment condition even to the least extent [1–7]. A lot of scientific knowledge has been incorporated nowadays into modern maintenance concepts in plants that has brought us today to a point where some sets of activities that were considered as the "job" of researchers and experts till a decade ago, is now considered as regular maintenance job. So, it is "just" an ordinary maintenance activity to check the condition of machines by performing ultrasonic testing, magnetic particle testing or dye-penetrant of cracks in equipment components, testing of microstructure changes in heavy metallic components (rollers, shell), vibrations analysis in frequency and time domain, analysis of oil in drives or plain bearings (even using scanning microscope), thermographic examination of heating and contact problems, statistical analysis of measured data and failures recorded, RFCA of process stoppages [1, 2, 8], etc.

Z. Stamboliska et al., *Proactive Condition Monitoring of Low-Speed Machines*,
DOI 10.1007/978-3-319-10494-2_1

Such applied "science" puts forward even more stretched targets in front of maintenance teams to find tools to extend the lifetime and have trouble-free operation of equipment. Tools of "classical" maintenance have been already well exploited and now researchers and advanced maintenance teams are focused on finding new techniques and methods to provide proactive, easy-to-incorporate condition monitoring methods that in small steps can take care of equipment condition and improve it, so that at the end-of-the-day, real lifetime will be beyond the expected, and stable and reliable operation will be provided [3, 9–12].

In relation to the current state of maintenance strategies, leading research centers in maintenance as well as worldwide industries related to maintenance products are clearly pointing toward trends in setting new key drivers related to achieving operational excellence, but at significantly lower costs, as a response to the latest economic crisis, redistribution of world economy growth, and expansion to the far East. Reflected in the area of maintenance, it is all about improving the organization's asset management, promoting new technical solutions, combining tasks, and enhancing the operator-driven maintenance strategies and proactive actions for extension of trouble-free equipment lifetime.

Extensive research and review have been carried out on existing methods for monitoring, measurement, and condition estimation in the literature, research in the World's centers of reliability and maintenance "excellence", and also recommendations by World's leading manufacturers of heavy-duty equipment, as well as specialized companies for maintenance solutions and vibrodiagnostics. The general policy of today's maintenance "World" of business considers maintenance as an integral part of all manufacturing processes. It is recognized as one of the main factors influencing overall costs of production, but also as a powerful tool for improvement of the overall efficiency of manufacturing and servicing industries. According to Roberts [13], the goal of today's maintenance is to provide a concept of full involvement of both "pure" maintenance and production operators for high overall reliability of equipment by employing the best available techniques and methods determined by benchmarking and coming close to the TQM concepts. Hutchins in [7] goes deeper and talks about two very distant, but core processes in today's successful maintenance strategies: the basic essential care of equipment and the strong empowerment of maintenance staff and operators for collection of any machine fault indications and proactive correction activities. The focus is on comprehensive planning, and as Crespo Marquez and Moreu de Leon write in [6], maintenance is today a complex management system where at first companies define a strategy and objectives derived from a business plan and then go to the technical level of maintenance methods, routines, data collections, evaluations, and corrections. Of course, the technical progress and new systems are a benefit enabled by today's easy available accurate measuring techniques, data collections, and analyzing systems [14–18] that provide new top records of plant machinery availability and run factors.

In some branches of industry, considered as "heavy-duty" industry, a lot of the facilities employed consist of many, so-called "low-speed" machines. They include cooling towers, low-speed agitators, paper machines, ball mills, crushers, dryers, rotary kilns, etc. Efficient operation and maintenance of these machines, the same as in other machines, is essential to maximizing production and minimizing downtime. However, the latent or incipient failures of these machines often go undetected due to people's mindset related to low-speed machinery as people could get used to the small incremental changes "forgetting" how the state was when the machine was totally free of faults. The main advantage of low-speed machinery that faults usually do not develop so fast into catastrophic failures, is also its main weak point because many "signs" go undetected considered to be a normal state. Therefore situations as cracks in housings, shells, trunnions, divot, or flat spot in low-speed bearings appear and it seems as "sudden" change, while the obvious signs were there all the time, in the machine. These difficulties are more expressed when having a combination of low-speed running machines and thermodynamic loads that could act quite fast. For example, rotary kiln in the cement industry runs at only 2–3 rpm and it is also considered as a robust machine. However, only a few minutes are required if something goes wrong with the brick lining inside the kiln to have total catastrophic failure of the kiln due to internal thermal energy characterized by temperatures ranging up to 1,400 °C.

Another, maybe even greater difficulty is that low-speed machines emit low impact energy when having some discrepancies in the moving component. This low energy also results in low vibration energy making the recognition of faults difficult in typical time or frequency spectrum measured. Moreover, if inappropriate measuring system is employed that is insensitive to such low frequency and low energy changes, any fault can be totally missed.

Parallel to this, very often these machines work in so-called nonstationary conditions with changeable loads, velocities, and extreme exploitation conditions, causing additional noise in the measured signals and, of course, difficulties to make correct interpretation of the results and diagnosis of machine health.

Hence, in low-speed machines the main problem we face is **how to accurately detect slow changes of low energy in equipment condition, very often in changeable exploitation conditions, and how to link them with the real root causes and potential failures rising in the stages where faults are still not completely developed**?

There are many known techniques developed at a very high level for condition monitoring and faults detection in medium- and high-speed machinery by analyzing mechanical vibrations (FFT, Cepstrum, Enveloping, etc.). Although there are no universally accepted criteria for classifying machinery as low-speed, intermediate, or high speed, it is generally recognized that low-speed machines are those that operate at or below 600 rpm [19, 20]; unfortunately, these techniques are not equally effective for monitoring faults of machines [19, 21, 22].

Broad research conducted reveals that there are much less sources on low-speed machinery maintenance and condition monitoring, and a majority of them [19, 20, 23–26] still deal with the speciality of low-speed monitoring techniques. Most of them are focused on application of new measurement and analyzing systems in the existing industry. For example, Mais in his work [20] explicitly points out that the first thing about low-speed machines is to understand the application and then to make a clear picture of how to incorporate vibration analysis (measurement system and data analysis), as low-speed machines induce low level of vibration.

The particular difficulty in monitoring low rpm vibration is that low frequency components are typically of lesser amplitude than higher frequency components [19, 20], which tend to drown out the low amplitude, low frequency vibrations of interest, producing unacceptably high signal-to-noise ratio. The problem for the monitoring system is also how to extract low level components (requires a wide dynamic range) from the composite signal.

Recently, ultrasonic techniques [27] are increasingly being used as they enable accurate detection of early damages to bearings with low speeds as ultrasonic probes can detect sound at about 40 kHz.

New approaches as "pattern recognition" for faults detection in complex machinery are applied and other various analyzing techniques [28, 29].

The other important note is that when dealing with low-speed machinery, it is of prime importance to understand the physical nature of the production process as its effects are often beyond the effects of the machine operation itself. For example, highly abrasive material rich in silica can cause extensive wear compared to the normal wear rate of linings in any crusher; inferior quality raw material could lead to destruction of protective coatings in cement rotary kiln and cause shell deformation or overfilling of any rotating equipment could cause destruction of sealing components, pollution of the working environment thus affecting moving components, etc. Although many such technological process effects are monitored in plants in some way, they are not used in a systematic way to provide early warning for possible failures and for undertaking preventive actions.

There are also valid international standards [30–34] that define the terms for condition monitoring, among which steady conditions of equipment are required in order to compare data on equivalent basis. However, research shows that for low-speed machines this is not entirely proper because conditions when faults are generating are omitted, which is actually characteristic of low-speed machines where change in mechanical components condition is more extensive mainly in unsteady conditions.

In addition to these technical issues is the fact that plant crew is limited and no one can afford the time as in the past to "take frequent walks" along the equipment. It is therefore of primary importance to bring the proactive and operators-driven maintenance on the shop-floor. Separate teams and maintenance centers within large Group companies are making efforts with own knowledge or specialized external companies to make the best of the application of the new maintenance trends in the industry.

In addition, in the heavy-duty industry very often key equipment is required to run continuously for 24 h during annual campaign. A single day stoppage may cost the company about 100,000 € or more considering lost production, labor, and spares for repair, and in addition, the spare may not be available and delivery time can be as long as or more than 10–14 months.

In regard to low-speed machine maintenance, there are today commercially available measurements where modern techniques are employed for specific measurements of the geometry of moving components, wear rates, eccentricity, deformations, strain, cracks, etc. Of course, these services are not cost-free and their price can be up to 10,000–15,000 € per major equipment. Also, the new trend is to make technical assessment of machines condition for estimation of overall health, and also to estimate the remaining lifetime. In such cases, often Fine Element Method modeling (FEM) is applied to make a model of the assessed machine in order to explore the most sensitive points and check them. It is recommended that such measurements be performed periodically for longer durations, such as once every 2 years or after every larger reconstruction or replacement of major equipment component. However, there are several drawbacks related to these measurements:

- The service is performed approximately once every 1–2 years and in most cases it can only "confirm" the existence of a fault. It is too late to prevent it.
- These services are expensive compared to the larger time period (two lines plant would cost 20–30,000 € per service, or 100,000–150,000 € in 10 years).
- In between the two sets of measurements, various faults may occur without giving warning to the operators and maintenance teams. Cases of destroyed bearing, deformations affecting large bearings, and contact surfaces destruction are only some examples.

Hence, these services are beneficial, but they do not act proactively for most faults and cannot provide the required reliability of equipment.

Another important aspect in condition monitoring of low-speed machines is to know the limits of monitored signals to be set and how the trend of approaching these limits can be recognized. Only performing monitoring on-site is not helpful enough from this aspect as users may be aware of certain changes, but they would be unaware of how close the equipment condition is to the failure limits. Also, very often some typical parameters, as vibrations, displacements, amp, etc., are monitored, while there may be some others that could be better descriptors of the condition. It is also important to locate monitoring equipment (gauges, sensors, etc.) to obtain reliable monitoring signals. It is here that in the last half decade that FEM simulations [35] have come into play and increasingly, these numerical methods are being used for simulation of purposes to discover some early warnings, advice on the type and locations of monitoring equipment, identify root causes of defects, and explain various phenomena present in the operation of complex systems. All these actions are made in a safe "virtual space" of the design software but with consideration of real boundary condition acting on numerical three-dimensional models of investigated objects.

Considering the above, as authors we were inspired to help bring about a new approach for condition monitoring of low-speed machines. Therefore, in this book there are numerous data and recommendations on how to select appropriate condition monitoring methods, measuring systems, and probes, how to organize proactive monitoring, and how to get early data from typical condition monitoring techniques to act proactively to reduce the potential for failures development. This approach was shaped as a method for proactive condition monitoring of one specific machine (cement rotary kiln), which was designed to fulfill the following:

- Enable to monitor continuously the equipment and recognize unsteady conditions when faults are generating.
- Give possibility for proactive action by giving online data to serve as a basis for actions to correct condition changes at all times during equipment operation.
- Be easy to be incorporated in existing control systems at low cost.

In general, the main benefit of having proactive condition monitoring in place is to provide early warning and the chance to react proactively on the equipment. The application of such monitoring provides also:

- obtaining in-depth knowledge on machinery trough FMCEA and determination of proper maintenance strategy by analyzing major failures modes;
- using FEM model simulation and estimation of observed symptoms on equipment condition to set correct limits for alarms of the condition monitoring;
- online follow-up of specific parameters keeping the operators updated for proper and on-time correction of the machines operation and avoid potentially dangerous situations.

Aiming to assist researchers on the practical side of maintenance, and to assist industry applicants in upgrading their own condition monitoring systems, this book contains the following:

- Review of current maintenance strategies and methods for condition monitoring with special focus on low-speed machines;
- Recommendations on how to determine the most appropriate maintenance strategy based on failure modes;
- Analysis of key differences and difficulties for condition monitoring of low-speed machines;
- Review of applied condition monitoring techniques for low-speed machines;
- Recommendations for the best practices for technical assessment of condition of most frequently applied low-speed machines;
- Description of proactive approach in condition monitoring and how to create such proactive monitoring system and understand the data we are getting in a way that these data will enable us act on time proactively;
- Use of FEM as beneficial integral technique for proactive condition monitoring;
- Case studies for applying the proactive method for condition monitoring of exemplary low-speed machine (cement rotary kiln) consisting of:
  - Determination of the most suitable real-time monitoring parameter for proactive condition monitoring
  - Design of measurement system in real exploitation conditions
  - Determination of methodology for estimation of equipment condition at all early initiations of possible faults

- Application of FEM to define limiting values for condition monitoring alarms
- Setting alarms as triggers for corrective actions in course of proactive maintenance of "healthy" equipment condition.

References

1. FLSmidth Institute: The International Maintenance Seminar, vol 1 (Copenhagen, 2005)
2. Polysius: Technology Forum (Neu Beckum, 2004)
3. R. Briš, E. Châtelet, F. Yalaoui, Optimization of maintenance cost under asymptotic reliability constraint, in *Safety and Reliability,* ed. by Bedford, van Gelder (© Swets & Zeitlinger, Lisse, 2003). ISBN:90 5809 551 7
4. M. Cepin, Optimization of testing and maintenance schedules, in *Safety and Reliability,* ed. by Bedford, van Gelder. (© Swets & Zeitlinger, Lisse, 2003). ISBN:90 5809 551 7
5. R.P. Chapman, *Recommended Procedures for Mechanical Analysis of Rotary Kilns* (Fuller Company, Bethlehem, 1985)
6. A. Crespo Marquez, P. Moreu de Leon, J.F. Gomez Fernandez, C. Parra Marquez, V. Gonzalez, The maintenance management framework: a practical view to maintenance management, in *Safety, Reliability and Risk Analysis,* ed. by Martorell et.al. Theory, Methods and Application (Taylor & Francis Group, London, 2009). ISBN:978-0-415-48513-5
7. H. David, *What is Total Productive Maintenance?* (David Hutchins International Quality College, Somerset)
8. Root Cause Analysis Guidance Document, DOE-NE-STD-1004-92 (Washington 1992)
9. M. Desinde, J.M. Flaus, S. Ploix, Risk analysis and diagnosis modelling for online control of process, in *Safety and Reliability for Managing Risk,* ed. by Guedes Soares, Zio (@ Taylor & Franics Group, London, 2006). ISBN:0-415-41620-5
10. E. Garbolino, J.-P. Chery, F. Guarnieri, Dynamical risk assessment in the process industries: proposal of a methodology based on the system dynamic approach, in *Reliability and Risk Analysis and Safety* (Taylor & Francis Group, London, 2010)
11. R.G. Herbert, D.J. Wallis, in *Are you Reaping the Benefits from your Monitoring System?*, EPRI International Maintenance Conference, Chicago, 19–20 Aug 2003
12. M.J. Kallen, J.M. van Noortwijk, Inspection and maintenance decisions based on imperfect inspections, in *Safety and Reliability* ed. by Bedford, van Gelder (© Swets & Zeitlinger, Lisse, 2003). ISBN:90 5809 551 7
13. R. Jack, in *Total Productive Maintenance (TPM)* (Technology Interface/Fall, USA, 1997)
14. B. Ale, C. Van Gulijk, P. Swuste, *Metamophors and Models, Reliability and Risk Analysis and Safety* (Taylor & Francis Group, London, 2010)
15. A.A. Andreani, E. Zio, F.K. Rodriguez, P. Viveros, Integrated analysis of system reliability and productive capacity of a production line, in *Reliability and Risk Analysis and Safety* (Taylor & Francis Group, London, 2010)
16. A. Barkov, A. Azovtsev, in *A New Generation of Condition Monitoring and Diagnostic Systems.* A collection of condition diagnostics papers on the Internet site. http://www.vibrotek.com/ref.htm
17. S. Dunn, in *Condition Monitoring in the 21st Century,* Assetivity Pty Ltd, The Plant Maintenance Resource Center, Revised: August 2009
18. T.A Dunton, in *An Introduction to Time Waveform Analysis* (Universal Technologies Inc, USA, 1999)
19. R.M. Barrett, in *Low Frequency Machinery Monitoring Measurement Considerations* (Wilcoxon Research, Inc)
20. Jason Mais, *Low Speed Analysis: Vibration Monitoring of Low-Rotational Speed Applications Using Accelerometers* (SKF Reliability Systems, USA, 2003)

21. Jason Mais, Scott Brady, *Introduction Guide to Vibration Monitorin: Measurements, Analysis and Terminology* (SKF Reliability Systems, USA, 2002)
22. D.H. Shreve, *Introduction to Vibration Technology, IRD Mechanalysis* (Inc, USA, 1994)
23. R. Klubnik, in *Measuring Displacement Using Accelerometers* (Wilcoxon Research, Washington)
24. Lifetime Reliability Solutions, in *Maintenance and Asset Management Strategy Using Physics of Failure Factors Analysis*, Internet source
25. Noria Corporation, in *Sleeve Bearing Lubrication-Machinery Lubrication*, Internet sources
26. A. Smulders, in *Time Domain Analysis* (AB SKF, Gothenburg, 2005)
27. Y. Kim, A.C. C. Tan, J. Mathew, B. Yang, in *Condition Monitoring of Low Speed Bearings: a Comparative Study of the Ultrasound Technique Versus Vibration Measurements*, WCEAM (2006)
28. A. Jablonski, T. Barszcz, Validation of vibration measurements for heavy duty machinery diagnostics. Mech. Syst. Signal Process. **38**, 248–263 (2013)
29. J.R. Ottewill, M. Orkisz, Condition monitoring of gearboxes using synchronously averaged electric motor signals. Mech. Syst. Signal Process. **38**, 482–498 (2013)
30. ISO17359 Condition monitoring and diagnostics of machines-General guidelines
31. ISO13372 Condition monitoring and diagnostics of machines-Vocabulary
32. ISO13373-1 Condition monitoring and diagnostics of machines-Vibration condition monitoring
33. ISO13374-3 Condition monitoring and diagnostics of machines-Data processing, communication and presentation-Part 3 Communication
34. ISO13379 Condition monitoring and diagnostics of machines-General guidelines on data interpretation and diagnostics techniques
35. E. Rusinski, S. Dragan, P. Moczko, D. Pietrusiak, Implementation of experimental method of determining modal characteristics of surface mining machinery in the modernization of the excavating unit. Arch. Civ. Mech. Eng. **12**(4), 471–476 (2012)

Maintenance Management and Applied Strategies

2

Abstract

This chapter describes the basics of maintenance management systems and key factors for successful organization and performance. It also points out the main characteristics of major maintenance strategies with their advantages and disadvantages.

Keywords

Maintenance strategy · Preventive maintenance · Predictive maintenance · Reliability-centered maintenance · Proactive maintenance · Maintenance system · Maintenance management · Objectives · Assets management · Human resources · Spare parts management · FMECA

2.1 Maintenance Management Systems

The European Federation of National Maintenance Societies defines maintenance as: "All actions which have as an objective to retain an item in or restore it to, a state in which it can perform the required function. The actions include the combination of all technical and corresponding administrative, managerial, and supervision actions."

Maintenance is an inevitable segment and essential activity in every manufacturing and service plant. Since the basic rule is to have a process that would gain more money than it costs, the only way to provide optimum reliability of equipment is to choose the most efficient and cost-effective maintenance for all components of the machinery. According to DuPont, maintenance is the largest single controllable expenditure in a plant and therefore it attracts the attention of the overall management system in the organization [1–3]. Following the many researches ongoing in the last decade, as Deloux et al. [4], Marquez [1] as well as Moubray [3], and many others [4–8], maintenance has gone through rapid change over the last century in order to develop strategies that would explicitly relate the system

Z. Stamboliska et al., *Proactive Condition Monitoring of Low-Speed Machines*,
DOI 10.1007/978-3-319-10494-2_2

performance to operating environment and create increased awareness that maintenance not only affects product quality, equipment availability, and costs, but also overall safety and environment.

Today, maintenance represents complex management processes within organizations. It associates many internal processes such as production, quality assurance, and environment improvement and in the last years, risk analysis in the organizations [4, 9] and safety of people and organizations. This process comprises:

- Definition of the maintenance strategy; and
- Implementation of the strategy [1].

The determination of the strategy is of prime importance and the starting point for any company. Below is given an insight into this item:

> Maintenance strategy represents a plan of actions or policies designed to achieve a targeted aim in a company. It is about integrating organizational activities and utilizing and allocating the limited resources within the organizational environment so as to meet the pre-defined objectives.

While planning a strategy it is essential to consider that decisions are going to affect the overall company's operations in both horizontal and vertical lines involving own employees, suppliers, and customers. It should provide conditions for effective maintenance (plans, schedules, controls, inspections, and improvements) resulting in minimized expenditure (direct and indirect costs) and company satisfaction with capacity and conditions of the assets.

Implementation of the strategy is related to the maintenance management ability to deal with many different implementation issues (finding properly skilled personnel, tools and monitoring systems and techniques, work monitoring, contractor management, etc.). However, implementation is much dependent on the definition of proper maintenance strategy and its appropriateness for the organization and equipment.

A good representation of today's good Maintenance management system was given by Crespo Marqez et al. in their work "The maintenance management framework: A practical view to maintenance management" [1]. An upgraded model based on this article is presented in Fig. 2.1, which describes the phases of the system and their role in maintenance effectiveness, efficiency, assessment, and continuous improvement.

The whole story starts actually from the overall company's targets and KPIs. The required efficiency of the processes is translated into tangible maintenance objectives and KPIs and their ranges that would provide the first.

In-depth knowledge is required of own equipment to make good prioritization of the assets and define their maintenance strategy. Tools as FMECA, RFCA, history records, manufacturers' recommendations are needed in this phase as they are of

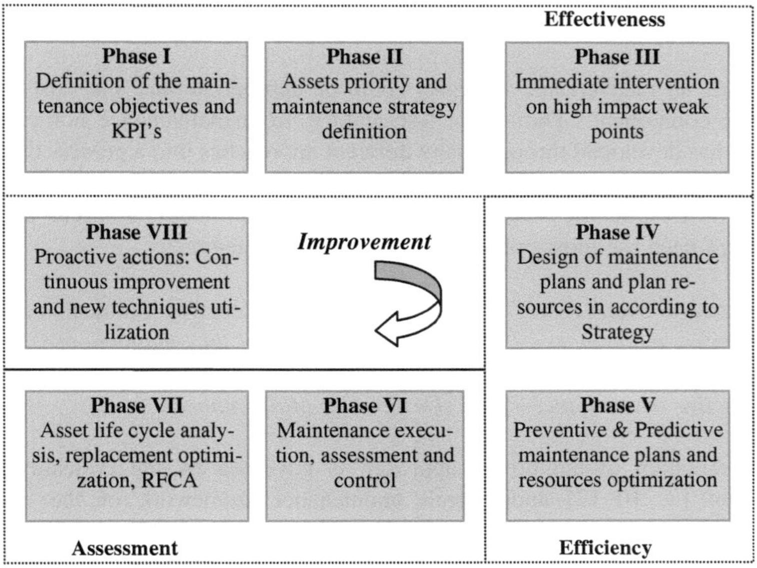

Fig. 2.1 Maintenance management model [1]

crucial importance. Recognizing the weakest points of the system or facility to be maintained, it may be necessary to intervene immediately on them with high impact to bring them to some level of reliability. The next phase is the operational phase where maintenance plans and needed resources are defined based on the maintenance strategy. Preventive and predictive maintenance plans are to be designed as fine-tuning of previous phase, schedules of inspections and actions, required tools, personnel, infrastructure, define predictive maintenance activities in the course of equipment lifetime optimization. This phase, or employed maintenance strategies, such as the preventive, predictive, proactive maintenance, reliability-centered maintenance are those that affect the efficiency of the maintenance overall. Having done this crucial preparatory part of maintenance, it is essential to execute maintenance activities as inspections, repairs, replacements, modifications etc., assessment of the same and control. It is the well-known PDAC principle applied: Plan-Do-Act-Check. Once these phases are ongoing, next is to analyze the assets' lifecycle and optimize replacement. Especially, the use of the RFCA in detection of the real root causes affecting the lifetime of an asset has been proven beneficial. This phase is also an input to the next phase representing the proactive approach, which is directed toward the improvement of overall maintenance by proactively undertaking various improvement actions, employing new techniques, etc.

The core issue in maintenance management process, as already mentioned, is accurately choosing the type of maintenance strategy to be applied for a certain equipment. To have a better understanding, the basic principles as well as advantages and disadvantages are given as follows.

2.2 Basic Principles of Maintenance Strategies

In general, all maintenance strategies could be appropriate and also cost-effective for some components. During the last century, from maintenance non-existence, this area has developed through many different approaches into a process that is one of the most significant processes in both production or servicing organizations (Fig. 2.2). Despite the stages of development, even today different pieces of equipment need the most optimized maintenance strategy:

- "*Operate to break down*" (or "fix when breaks"-Reactive Maintenance: ReM),
- *Preventive maintenance-PM* (or some called *Fixed time maintenance),*
- *Predictive PdM (Condition based maintenance: CBM)* or most sophisticated
- *Pro-active maintenance-ProM (Design Out Maintenance).*

Determination of the proper approach is governed by the criticality of the equipment [3, 10–12] and overall maintenance framework of the company (Fig. 2.2).

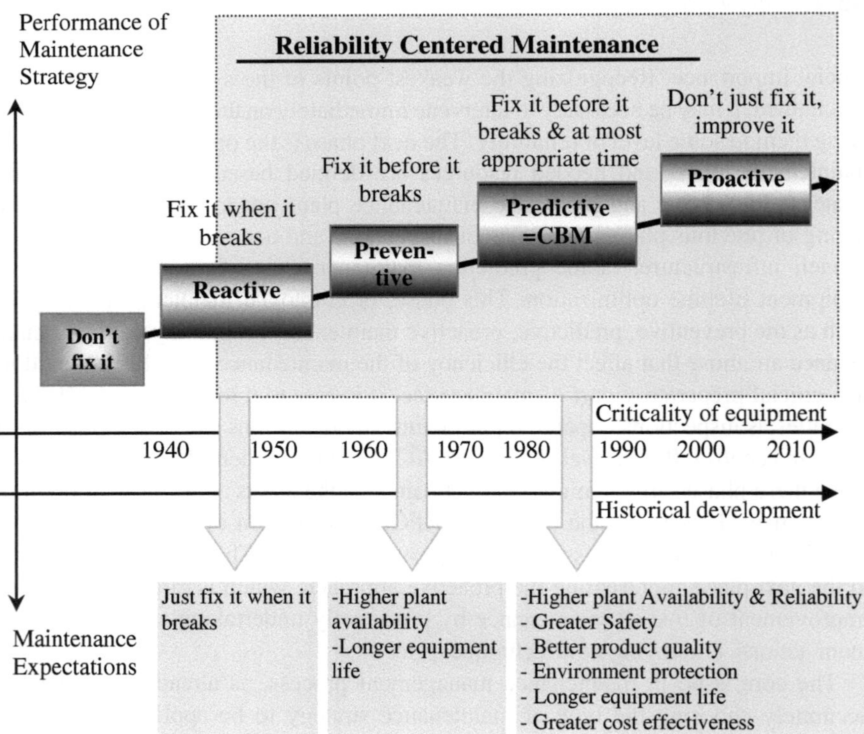

Fig. 2.2 Main maintenance strategies [3, 10]

Maintenance strategies have started being developed from the basic reactive approach, where the maintenance teams were huge and focused everyday on the correction of existing defects dealing with large amount of spares and no management of any maintenance issue.

Becoming aware of the meaning of equipment reliability that was providing continual throughput of the industrial lines with less quality difficulties, companies and researchers have started organizing systematic planning and scheduling of repair activities involving technologies for condition monitoring of equipment. Thus maintenance teams were able to fix problems before a failure would happen. However, this was still not enough in modern competitive industries where the production capacities were maximalistic and every downtime was a risk to lose potential client as well as causing profit/loss due to not actualized production. It was inevitable to continue efforts in further development of maintenance strategies and the predictive and later proactive approach was the answer. Technologies of condition monitoring were upgraded with statistical methods of data processing, use of more sophisticated methods, and probability methods for estimation of equipment components' lifetime. This led us to a state where potential failures could be sensed, lifetime predicted fairly accurately, and maintenance was able to undertake corrective actions in planned shutdowns, thus not disturbing the production and even more, inventory could to be better planned and optimized. In conditions of tough competition, but even more, in the recent economical crisis, in the first decade of this century, every possible savings in production was welcomed and pressure was put on all involved parties to improve both fixed and variable costs. Time difference between planned outages was challenged to increase and it affected maintenance further to work not only at correcting, but also at improving equipment and to extend their lifetime. Therefore proactive maintenance has taken the leading role in most modern industries.

Today, the meaning of the modern maintenance organization is founded on anticipation *what will happen in the future* and *planning and scheduling corrective and improvement actions in advance* [13, 14]. It is "*Thinking*"-oriented approach where the first step that companies tend to follow is to change the work attitudes or awareness. It is actually a complex maintenance system called ***Reliability-Centered Maintenance (RCM)*** comprising all activities of the mentioned formalized maintenance strategies (ReM, PM, PdM, PrM) and applying them on the overall criticality and risk assessment of the organization and plant equipment.

2.2.1 Preventive Maintenance

The activities for taking care and servicing for the purpose of maintaining equipment and facilities in satisfactory operating condition by systematic inspection, detection, and correction of incipient failures either before they occur or before they develop into major defects is known as preventive maintenance [3, 10]. According to ISO13372: 2004 (E), it is maintenance performed according to a fixed schedule, or according to a prescribed criterion that detects or prevents degradation of a functional structure, system, or component in order to sustain or extend the useful life. It is a schedule of planned maintenance actions aimed at the prevention of breakdowns and failures thus enhancing equipment reliability.

Preventive maintenance activities include so-called Essential Care activities (cleaning, lubrication, alignment, adjustments, filtration, balancing, and good operations practices) and partial or complete overhauls at specified periods (Fixed Time Maintenance). In addition, for good and effective preventive maintenance, it is necessary to have well-organized and documented system scheduling periodical checks, corrective actions, and evaluation of findings and incorporating them into the maintenance activities. Follow-up of all findings and corrective actions taken are of prime importance for achieving full benefit of the preventive maintenance. Commercially available softwares (as SAP, SKF Marlin, CMMS PMXpert, Fast-Maint CMMS, EVAM- SPM etc.) exist to support the organization in preventive maintenance.

There are multiple misconceptions about preventive maintenance such as that it is unduly costly, and this was already argued by many authors such as Castro, Huynh, Barros, Berenger, Pantazopulos, etc., in [10, 14, 15]. This logic dictates that it would cost more for regularly scheduled downtime and maintenance than it would normally cost to operate equipment until repair is absolutely necessary. What here is often elapsed is that such "absolutely necessary repair" can be needed in very unfavorable times for the market/production and also in most cases there will be already developed other significant damages of related components tending to fail soon after.

Long-term benefits of preventive maintenance include:

- Improved system reliability.
- Decreased cost of replacement.
- Decreased system downtime.
- Better spares inventory management than of corrective maintenance where all items are required at all times.

2.2.2 Predictive Maintenance

Predictive maintenance techniques help in determination of the condition of in-service equipment in order to predict when maintenance should be performed. This means it does not prevent anything itself, but gives information on failures that are developing toward a breakdown. This gives plants the possibility to plan and schedule maintenance corrective actions and provides all necessary resources. According to ISO13372:2004, it is maintenance strategy that emphasizes prediction of a failure and taking action based on the condition of the equipment to prevent failure or degradation. This approach offers cost savings over routine or time-based preventive maintenance because tasks are performed only when warranted. It can be performed within the planned maintenance shutdown, or in the best case even use operational shutdown such as setup, cleaning, process changes, market conditions, etc.

The "predictive" component of the term Predictive Maintenance stems (PdM) from the goal of predicting the future trend of the equipment's condition. This approach uses principles of statistical process control to determine at what point in the future maintenance activities will be appropriate. For example, as shown by many condition-monitoring techniques [7, 16–21] using the vibration monitoring of a bearing, by PM it would be recognized that the bearing has damage and it would have to be replaced. Unlike PM, with PdM, mathematical trending could be used as tool to estimate based on the historical progress of the bearing condition parameters and working conditions when a failure could be expected, early enough to organize corrective actions.

Most PdM inspections are performed while equipment is in service, thereby minimizing disruption of normal system operations.

In the literature and among industry users, often Preventive maintenance is called a maintenance system containing all activities of condition monitoring, misconceptually called as Predictive maintenance. Based on the opinion of other resources [3, 12] and the authors of this book, Predictive maintenance system is a maintenance strategy that actually has been superstructured over Preventive maintenance. It is a system that uses the tools of condition monitoring as a comprehensive way to determine possible failures early enough and then act to prevent potential failures. It has been also historically developed after the establishment of preventive maintenance, but it is actually a kind of "controlling management system" for the actions of preventive maintenance.

PdM is also called **condition based maintenance** as it attempts to evaluate the condition of equipment by performing periodic or continuous (online) equipment monitoring. The ultimate goal of PdM is to perform maintenance at a scheduled point in time when the maintenance activity is most cost-effective and before the equipment fails in service. However, it must be pointed out that in most cases the maintenance management decides in establishing combined systems for preventive/predictive maintenance system based on the criticality of equipment and costs required to have the system running.

2.2.3 Proactive Maintenance

It is an industrial improvement approach focused on identifying and establishing the operational, maintenance, and capital improvement policies together that will manage the risks of equipment failure most effectively. According to ISO13372:2004, it is a type of maintenance emphasizing the routine detection and correction of root causes that would otherwise lead to a failure. Accepted methods of PM and PdM strategies combat machine damage based on either detecting the warning signs of failure once they have already begun (predictive) or regular maintenance according to a schedule rather than the machine's true condition (preventive). None of these take a micro view on machine damage—concentrating on the causes instead of the symptoms of damage.

Proactive maintenance commissions corrective actions aimed at the sources of failure. It is designed to extend the life of equipment components as opposed to (1) making repairs when often nothing is broken, (2) accommodating failure as routine and normal, and (3) preempting crisis failure maintenance—all of which are characteristics of the predictive/preventive disciplines. Proactive Maintenance is all the techniques that we can use while equipment is functioning well. It actually emphasizes the use of Predictive Maintenance techniques in addition to traditional preventive measures. Based on Proactive maintenance, three main risks from equipment failures are to be recognized:

- the risk to safety and environment,
- the risk to operations, and
- the risk to the maintenance budget.

Based on these risks, proactive maintenance applies any one of the known maintenance strategies:

- on-condition maintenance tasks (CBM),
- scheduled restoration or discard maintenance tasks,
- failure-finding maintenance tasks, and
- one-time changes to the "system" (changes to hardware design, to operations, or to other things) so that a prolonged lifetime would be obtained.

According to the analysis of Moubray [3], some of the most beneficial improvements that proactive maintenance has introduced are the following:

- changing from efforts to predict life expectancies to trying to manage the process of failure,
- an understanding that the vast majority of failures are not necessarily linked to the age of the asset,
- an understanding of the difference between the requirements of an assets from a user perspective, and the design reliability of the asset,
- an understanding of the importance of managing assets on condition (often referred to as condition monitoring, condition-based maintenance, and predictive maintenance),
- an understanding of the four basic routine maintenance tasks [*on-condition maintenance tasks, scheduled restoration or discard maintenance tasks, failure-finding maintenance tasks, and one-time changes to the "system" (changes to hardware design, to operations, or to other things*)], and
- linking levels of tolerable risk to maintenance strategy development.

Proactive maintenance in the industry has been mainly developed in the areas of business where production capacities are at upper limits, market is favorable for the maximalistic approach, and overall management is at a level understanding the significance of the equipment reliability and its influence on the production throughput and quality. In these cases, the maintenance teams promote every improvement steaming not only consistent equipment operation, but also extending the lifetime of components. This means that technologies of condition monitoring are in regular use, but the findings are applied to "find a better way" to maintain. It is typical to employ new materials for abrasion-exposed equipment, apply special coatings made of ceramic or similar material to promote abrasion resistance, apply new types of greases with better consistency, wash-out and carrying properties, etc. However, proactive maintenance is not only "material" maintenance set of activities. It is also a managerial approach where some of the maintenance basic tasks are transferred to the operators to get a driving force and continuous feedback on the actual and day-to-day state of the equipment in operation. Understanding the basic principles of the equipment function and possible failures, proactive measures are undertaken even at this level of operators. It is no longer a facility run to get a quality product on time, but it also has facilities for equipment condition care. Typical examples are operators focused on machinery amps, provision of operating conditions that will not distort equipment condition (especially important in thermodynamic systems), keeping the equipment running in the optimized condition (set point of load, velocities, temperature, etc.).

2.2.4 Reliability Centered Maintenance

It is maintenance system based on comprehensive consideration of what must be done to ensure that any physical asset continues to fulfill its function in the present operating context. According to ISO13372:2004 (E) it is disciplined logic used to identify those cost-effective and technologically feasible maintenance tasks that realize the inherent reliability of equipment at a minimum expenditure of resources over the life of the equipment. The term Reliability-Centered Maintenance (RCM) was first used in public papers authored by Tom Matteson, Stanley Nowlan, Howard Heap [22], and other senior executives and engineers at United Airlines (UAL) to describe a process used to determine the optimum maintenance requirements for aircraft. It is defined by the technical standard SAE JA1031, Evaluation Criteria for RCM Processes. The RCM starts with the seven questions given below, well elaborated by Moubray [3] and are listed as follows:

1. What is the item supposed to do and its associated performance standards?
2. In what ways can it fail to provide the required functions?
3. What are the events that cause each failure?
4. What happens when each failure occurs?
5. In what way does each failure matter?

6. What systematic task can be performed proactively to prevent, or to diminish to a satisfactory degree, the consequences of the failure?
7. What must be done if a suitable preventive task cannot be found?

RCM is an engineering framework that enables the definition of a complete maintenance regime enabling machinery stakeholders to monitor, assess, predict, and generally understand the working of their physical assets. This is embodied in the initial part of the RCM process, which is to identify the operating context of the machinery and write a Failure Mode Effects and Criticality Analysis (FMECA) [2, 23]. The second part of the analysis is to apply the "RCM logic," which helps determine the appropriate maintenance tasks for the identified failure modes in the FMECA. Once the logic is complete for all elements in the FMECA, the resulting list of maintenance is "packaged," so that the periodicities of the tasks are rationalized to be called up in work packages [3, 24–26].

RCM recognizes three major categories of maintenance actions as follows:

- *failure-finding.* Failure-finding tasks entail checking hidden functions periodically to determine whether they have failed (whereas condition-based tasks-predictive maintenance entail checking if something is failing).
- *redesign.* Redesign entails making any one-off change to the built-in capability of a system. This includes modifications to the hardware and also covers once-off changes to procedures (proactive maintenance).
- *no scheduled maintenance.* As the name implies, this default entails making no effort to anticipate or prevent failure modes to which it is applied, and so those failures are simply allowed to occur and then repaired. This default is also called *run-to-failure*.

This approach means that proactive tasks are only specified for failures which really need them, which in turn lead to substantial reductions in routine workloads.

2.2.5 Total Productive Maintenance

Total productive maintenance (TPM) is more of a maintenance concept than a real strategy, but in many sources it can be found as such. It is based on the joint responsibility of supervisors, operators, and maintenance staff to provide machines operating smoothly and also extend and optimize their overall performance. Dr. Jack Roberts from TAMU-Commerce in his work [26] points to TPM as a reassemble of TQM in several aspects, such as (1) total commitment to the program by upper level management is required, (2) employees must be empowered to initiate corrective action, and (3) a long-range outlook must be accepted as TPM may take a year or more to implement and is an ongoing process. The most important aspect is providing convinced people and management to TPM success. It uses the tools of the previously described maintenance strategies, but the most important of the TPM is actually education, comparative processes of maintenance techniques and methods, "benchmarking" and involvement of operators for daily maintenance routines, adjustments, lubrications, etc. The main goals of TPM are measured by the [27]:

- Overall Equipment Effectiveness (OEE),
- Performance Rate, and
- Quality Rate.

Following the Japanese experience [27, 28], TPM has actually reduced the need for outsourcing of part of production lines due to its operational effects. It is nowadays widely used in many international manufacturing companies more as a philosophical concept of maintenance that incorporates all the needed techniques from PM, PdM, and PrM.

2.3 Structure of the Maintenance Management System

When considering any kind of machinery and defining its maintenance management system, both function and performance standards for the machinery should be determined based on two main aspects: the primary and the secondary functions [1, 3, 15]; and even more, all of them have to be considered in the light of three main categories of influencing factors such as the internal environment, human factor, and external environment.

Luyk and Rouvroye have given one of the best descriptions in [5] according to which the functions of a machinery describe the following properties and performances.

Primary functions: functions for which the asset has been supplied and are described with data such as output, carrying or storage capacity, product quality, speed, customer service, etc.

Secondary functions: it is about functions that asset is expected to do more than simply fulfill the primary functions. These functions are safety, compliance with environmental regulations and impacts, comfort, economy, structural integrity, efficiency of operation, sometimes appearance of the asset, etc.

The three categories of influencing factors, mentioned above, are focused on

- Internal environment: technology (technological process, interaction with other internal processes as production lines, maintenance, customers, etc.) and organization (structure of the organization, culture and strategy of the organization).
- Human factor: consideration of the individual level, degree of direct interaction of the human to the process and systems, etc.
- External environment: factors of the interface between the organization (company, plant) and external environment (stakeholders).

When determining maintenance strategy it is necessary to consider one complex set of action areas and interrelate them toward the achievement of the final target.

In accordance to this, the key structural areas of every maintenance management system to be determined are as follows:

- Maintenance objectives in line with business objectives
- Assets management
- Human resources
- Spare parts management
- Determination of type of maintenance (strategy)per equipment
- Performance measurements and improvement

Unlike in new plants, when everything is being established from zero ground, for existing plants the best approach is to start with a targeted maintenance audit involving limited number of people with experience from the plant.

In the literature, information can be found on how to perform an audit, but the basic rule is to make a realistic picture of the current state of the plant condition and maintenance functionality in order to map the weak points, see the systems in place, and decide on the required improvement.

2.3.1 Maintenance Objectives

When determining maintenance strategy of an asset or of the whole facility, we have to start from the maintenance objectives and their management through solid KPIs. In parallel, we know that the work done by maintenance needs to support the business aims and operating strategy, so the correct way is to link the maintenance performance to the company's business (see Fig. 2.3).

Creating pathways from the top to the bottom of the organization, we actually determine the previously mentioned primary and secondary functions and all influencing factors. The next thing is to bring those items on a realistic basis and decide on the most effective and efficient KPIs per department and to individuals, as they are actually the core causes with effects to the overall company operations (looking from plant aspect).

There are many different parameters to be estimated and some of them are given in Table 2.1 categorized by the base property of maintenance they are portraying.

What is important is that an organization should not be drawn down in too many KPIs and too many parameters, but focus on the main ones for the company's business and policy. So, when establishing a maintenance management system and setting up strategies, this is the basic point to start with. Therefore, a lot of attention

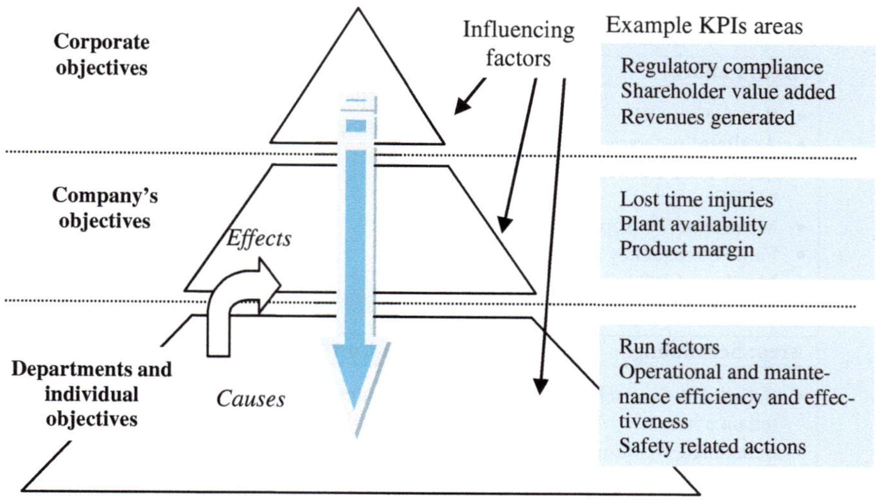

Fig. 2.3 Interrelation of company's objectives KPIs [29]

should be paid to what we want to achieve (what KPIs and what values) and set loops for control of the performance.

Having the KPIs determined based on the production and maintenance objectives, it is clear what the organization of the maintenance management system should be like and what strategies should be incorporated. It is clear that the company going upward with fully sold-out production capacities would strive to top reliability of equipment, so the best maintenance practices and predictive/preventive maintenance strategies would be economically justified to be employed. On the other hand, a company in the not so favourable position on the market would focus more on the relation of equipment reliability-to-costs reduction and maintenance strategies would surely look different.

2.3.2 Assets Management

Once we have determined our objectives on maintenance (for example high run factors, low equipment downtime, decrease of maintenance costs per unit product, etc.), we should focus on the assets of the plant. Often there is misunderstanding about what is considered under an asset and what asset management consists of.

> Any physical or intellectual property that is of any value to a company is called an asset. When referring to maintenance, it is about:
>
> - operational equipment for the production and supporting processes,
> - tools and equipment used for maintenance of the plant,
> - software and systems for maintenance, and
> - intellectual skills and know-how of the personnel.

Table 2.1 Typical maintenance-related KPIs [12, 30, 31]

Area: Failures	**Area: Training**
• Run factors (RF) • Availability factor (AF) • Mean Time Between Failure MTBF • MeanTime To Repair MTR • Failure Frequency / Failure Rate • Number of RCAs, Weibull analysis, PARETO analysis	• Skills matrix completion % • Personnel training hours
Area: Scheduling	**Area: Costs**
• Scheduled vs unscheduled maintenance hours • Scheduled Working Orders (WO) complete within planned estimate	• Maintenance cost ((materials, services, labour) per unit output • Craft utilization (hrs. reported vs. actual hrs.)
Area: Operational	**Area: Maintenance strategy performance**
• Total hours spend on maintenance per unit of product • Percentage of man hours/Work Orders of PdM activities from total hours/Work Orders. • Percentage of man hours/Work Orders of corrective WOs due to detection from PdM activities. (Vibration analysis, Infrared / Ultrasound surveys, Tribology) • % Overtime hours.	• Number of assets with reliability trending up • Numbers/hours of over maintenance activities reduced from budget

When determining maintenance strategy, plants must make a detailed list of all operational and other equipment related to production and supporting processes. A database should be created consisting of the following information:

- Type and technical properties of the equipment (capacity, power rates, speed…);
- Role of the equipment for the overall operations (function description);
- Breakdown analysis of machines into main assemblies down to unit spare parts;
- Technical drawings for equipment;
- Manuals (operation and maintenance);
- Lubrication requirements,
- History records of breakdowns, modifications, replacements, etc.; and
- As-build documentations.

The last, but not the least important step is the determination of equipment criticality. This is actually the basis for determination of type of maintenance to be applied, or what facility's resources, engineering effort, operations practices, maintenance, and training are to be employed to provide the item's continued operation. Defining equipment criticality is developing an equipment risk profile. Many sources can be found in the literature, but one of the simplest and still most effective method is to estimate criticality by use of the so-called risk rating indicator:

$$\text{Equipment Criticality} = \text{Failure Frequency}(/\text{yr}) \times \text{Cost Consequence}(€) \\ = \text{Risk } (€/\text{yr})$$

- "cost consequence" is the cost of lost production plus the cost of repair;
- "failure frequency" is from the company's maintenance history, or industry norms for a similar situation.

Equipment that stops production, or that causes major production costs when failed, is considered most critical.

When defining criticality of a plant's equipment, the process should be supported by a competent team of people consisting of operators, maintainers, and designers of the plant who contribute with their knowledge and experience. The team reviews documentation of the facility's processes and equipment. Equipment by equipment they analyze the consequences of failure to the operation and develop a table showing each equipments criticality rating.

There are various templates to be found for defining criticality of a piece of equipment and readers are directed to make review and use some of the existing or make one ones.

2.3.3 Human Resources in Maintenance

A plant may have state-of-the-art equipment for maintenance purposes (condition monitoring devices, tools, process monitoring software, etc.), but still have poor effectiveness and efficiency of maintenance.

> Equipment may fail unexpectedly, despite being officially the subject of preventive/predictive maintenance. Where is the clue then?-this is a frequently asked question in companies' upper management circles. The answer is at the human resources, in particularly their engagement.

Not to sound too philosophically, human resources in maintenance should be organized from the following aspects:

- Have optimized clear systematization for the plant needs with several, but not too many levels of responsibilities.

- Have clearly defined responsibilities of the personnel and match the people personal skills to the positions.
- Have specific person/team responsible for implementation and development of condition monitoring with strong position and reporting directly to top maintenance level. This team should be empowered and truly believe in the importance and benefits from the condition monitoring, but also develop skills for prediction and proactivity.
- Develop strong working culture, especially related to thoroughness of the repairs, planning skills, reduction of idle time, and control of the work toward keeping complete integrity of the equipment.
- Training and development of maintenance-related skills and know-how as well as regular implementation of applicative novelties related to maintenance aspects.
- Bring the various specialities into one team (e.g., mechanical, electrical, and civil maintenance to feel as one team).
- Have or develop skilful contractors.
- Have good cooperation with operations personnel.
- Bring the spirit of full engagement of personnel as they can make the plant efficient and reduce costs significantly, or can make the plant very inefficient with plenty of stoppages, overtime, "difficult life" of maintenance, and finally cause injuries due to equipment malfunctioning or exhaust work.

2.3.4 Spare Parts Management

The purpose of spare parts management is to ensure that right spare part and resources are at the right place (where the broken part is) at the right time. It is a relatively complex management system because:

- the spare parts might never be used,
- the parts might not be stored properly, leading to defects after installation,
- maintaining inventory of spare parts has associated costs, and
- parts planned to be obtained by contracted supplier may not be available when needed even from the supplier.

Spare parts management is an important economical issue and sometimes companies can create huge inventories trying to have available all required spare parts, thus keeping part of their potential profit idle. An effective spare parts management consists of optimized supply sequence that provides spare parts on time, optimizes inventory, disposes waste in a safe way for the environment, and reduces overall supply costs.

When starting new plants, it is important to make initial stock of the most critical spare parts that may lead to long stoppages of the production line. Therefore, right after commissioning, a team should be formed to focus on the manuals and recommendations of the manufacturers for spare parts of the critical equipment. After making a list for the various equipment, spares should be analyzed and matched, e.g., the same bearings may be installed in a few different machines. The best practices show that smartly chosen codification of the parts and software support help to create a unique database with various filters in order to find common spare parts and determine optimized stock. Of course, there will be strategic and unique spare parts that will have to be supplied even for a single machine.

At existing plants, the same as new plants, very important questions are:

- Should a spare part be at all supplied on stock?
- What is the right number of items to be kept on stock?

When deciding if a spare part should be supplied, the following questions should be considered:

- Is the equipment crucial for the production line?
- Can the part be repaired in case of malfunctioning?
- Is it possible to use similar spare part with some in-house modification?
- Is the cost of repair justifiable in relation to new part and costs of lost production?
- Can the part be made locally (what quality, costs, time, lost production)?
- Can the part be kept in supplier's inventory (arranged with some special contract)?
- In case of more capital spare parts, what is delivery time?

Different processes should be in place to manage **critical parts**—parts that a production relies on, and **consumable parts**—parts that are not critical to the line (like bolts and nuts, simple seals).

According to the Reverse Logistics Association, the primary focus of inventory control should be on the active and most critical items.

Helpful support in this case is to have a well-organized supply chain and keeping some commonly used spares on stock at the supplier (like bearings, seals, consumables, smaller electromotor, metal materials, etc.)

The second question about what is the right number of items to keep on stock is related to detailed analysis of multipurpose use of spares and frequency of replacement. It is essential to analyze the spare parts inventory based on various characteristics such as frequency of use, annual consumption value, criticality, lead time, and unit price. The commonly used inventory analyses are:

(1) FSN Analysis (based on frequency of issues/use)
(2) ABC Analysis (based on the lead time)
(3) VED Analysis (based on criticality)
(4) SDE Analysis (based on consumption)
(5) HML Analysis (based on unit price)

More details can be found in the literature, so the readers are encouraged to find and use the one most appropriate.

2.3.5 Determination of Maintenance (Strategy) Per Equipment

The next phase in the maintenance management system, after assets have been set with their priorities [5, 30] is to define the most suitable maintenance strategy to be able to achieve defined KPIs. It is important to understand that the best approach, as included in RCM, is to analyze asset by asset and determine the most suitable and optimal maintenance strategy per asset. Attention should be paid to assets with high criticality and of course not all assets in a production line should be subjected to predictive maintenance for example. For some of them it is enough to correct them when failing as they do not influence any of the main factors: Safety, Environment or Production, and in addition, they can be fixed easily at low cost and without line stoppage.

In the literature, different approaches can be found on how to determine maintenance strategy of an asset. In general, they are all based on the criticality of the items, effects of their operation and failures, costs, availability of resources, etc. The aim of this book is not to overview these approaches, but to focus on one of the most effective and systematic approaches to determine a maintenance strategy by using Failure Modes Effects and Criticality Analysis (FMECA) technique. This technique is widely used for analysis of assets and the procedure is given in detail in the valid standards and other literature sources.

> But in this book, this technique is used in different way as to analyze the failure modes themselves in order to determine the priorities for condition monitoring techniques of separate failure modes and appropriateness for application of certain maintenance strategy. In this way, we don't use FMECA for determination of a maintenance strategy as one for the whole one asset, but it is used to determine strategy for certain property of the asset.

Before explaining this new use of the technique, some of the basics steps of this analysis are given in the next sub-chapter together with an example of a heavy-duty industrial equipment that will be later used also as example for the other steps of the new method for proactive condition monitoring.

2.3.5.1 Analysis of the System to Be Maintained

Failure mode effect and criticality analysis (FMECA) as defined in [22] is an extension to FMEA procedure for analysis of a system to identify potential failure modes, their causes, and effects on system performance including means of ranking the severity of the failure modes to allow prioritization of countermeasures. It is usually done by combining the severity measure and frequency of occurrence to produce a metric called criticality. One of the main general purposes of applying this analysis is exactly to allow improvements of system's maintainability by highlighting areas of risk or nonconformity for maintainability. It also helps to identify failures having major unwanted effects on the system operation.

In order to build a proper maintenance strategy for the core equipment, detailed insight is required into major equipment failure modes that could affect its normal operation and reliability and review possibilities for their prevention and correction in terms of feasibility, costs, and time. FMECA enables determination of the criticality or priority for addressing certain failure mode with respect to the system's correct function or performance, but also classification of the identified failure modes according to their ease of detection, which is directly linked to the condition monitoring techniques used.

FMECA should be conducted following the required stages [22]:

1. Identification of failure modes
2. Determination of their respective causes
3. Recognition of effects: (consequences in terms of operation, function or status)
    - local effects (effect on the machine's component under consideration); and
    - final effects (impact on the highest system level)
4. Determination of detection methods or ways in which the failure can be detected and the means by which the users or maintainer is made aware of the failure;
5. Classification of severity of the failure mode's effect on the machine or component operation considering the effects to the users or environment, functional performance, safety requirements, contractual requirements, etc.;
6. Determination of frequency or probability of occurrence of each failure mode in order to adequately assess criticality of the failure mode.

Items under 4 (detection), 5 (severity), and 6 (probability of occurrence) are usually considered together under the so-called "criticality" or Risk Priority Number (RPN) used for prioritization in addressing the mitigation of failure modes.

$$\mathrm{RPN} = \mathrm{S} \times \mathrm{O} \times \mathrm{D}$$

where:

- *S*—severity grade;
- *O*—probability of occurrence;
- *D*—detection, or estimated chance to identify and eliminate failure before the machinery is affected (higher score means higher difficulty to detect)

For the purposes of this book, an example of FMECA for cement rotary kiln is given (see Table 2.2), without elaborating all previously given steps and used charts and criteria. Herewith, the failure modes, causes and effects are analysed as well as final rating of the RPN for each failure mode.

For all these failures to happen, there are different direct causes that are actually representation of a state of some of the kiln components. The real **physical root**

Table 2.2 FMECA of cement rotary kiln [2, 11, 32–34]

	Failure mode	Failure local effects	Failure final effect	Severity	Cause	Occurrence scoring	Available controls and condition monitoring methods	Detection	Risk Priority Number RPN
Kiln unavailability or reduced production capability	KILN AXIS DISTORTION	Overloading of stations and rollers causing hot bearings, cracks in the roller or shell	Kiln failure -total inavalibaility due to cracks in shell, rollers	9	Wrong assembly of tyres, rolles or even kiln shell sections or Incorrect cooling and heating up procedure or Tyre-roller wearing	6	Geometric measurement in hot condition (POLSCAN, FLS, Phillips)	6	324
	KILN AXIAL MISBALANCE	Disability for kiln axial travel, hot thrust bearing, increased wearing of tyres/rolers	Possible kiln stop (few days) due to bearing seizure (hot thrust bearing)	7	Uneven coatings, Assembly of new tyres or rollers, Shell "hot"spots, Thermal overexpansion	7	At hydraulic thrust rollers, it is controlled via hydraulic pressure. At classical kiln no objective control.	8	392
	KILN CRANK	Overload of rollers and bearings, Contact failure resulting in huge impact loads, overload of kiln drive	Kiln failure -total inavalibaility due to cracks in shell, rollers, hot bearing	8	Uneven coatings arround circumference, Longitudinal uneven coatings (cantilever effect), Thermal overexpansion, Incorrect cooling or heating up, "Hot" spots in the shell	7	Only subjective (visual) control that is not always applicable and it could be too late for corrective actions	8	448
	KILN OVALITY	Shell deformations causing difficulties at brick lining and their life time	Loss of linings or Kiln shell cracks -kiln inavalibility of few days	7	Thermal overexpansion	6	Kin shell scanners and migration monitoring	2	84
	BEARING BOUNDARY LUBRICATION	Hot bearing reducing kiln capacity for few hours at least	Kiln inability of hours or few days (hot plain bearing, damaged Babbit bearing or crack in roller)	7	Wrong bearing assembly, Failure of lubrication system, Overloading of bearing or Incorrect kiln cooling or heating up procedure	7	Temperature probes in the bearings (this indication can too late for remedies)	8	392

Note The scoring is from 1 to 10 (1 is lowest, 10 is highest score)

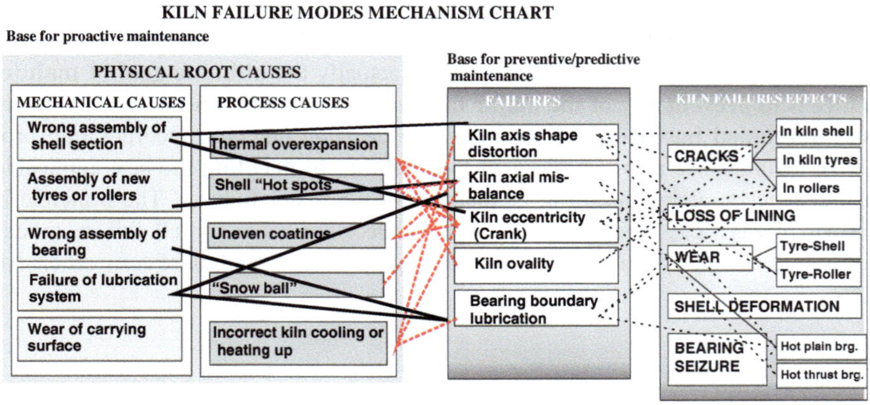

Fig. 2.4 Interrelations among kiln failures and their root causes [33]

causes (here the term "physical" has been used because in separate cases of failures, the final root cause may be also organizational or human error, but for this purposes only the physical nature of the failure is considered). The relations between the kiln main failure modes, and the real root causes are represented on Fig. 2.4. The causes shall be analysed later, but it is of crucial importance to understand the complexity of the interrelations among different failure modes, effects, and causes.

As shown in Fig. 2.4, the **failure modes** are actually several critical items defined as state of some of the kiln components. In order to prevent any of the faults to develop into real kiln failure, these states should be monitored with selected methods, but it should be pointed that monitoring of these items is predictive/preventive strategy that registers already actual state and it can prevent only the final stage of the failure. A **proactive condition monitoring would be actually monitoring of the initiating phases of the so-called "Physical root causes"** shown at the left-hand side of the chart.

To summarize, use of FMECA can help understand the system better and decide on what are the most critical failures in order to address them in more details. It also helps to determine the root causes of the failures. With this information, analysis can go further to check whether root causes can be monitored and recognized at their very beginning and if any preventive measure is possible. This is the input to the next stage of determining the maintenance strategy of the equipment.

2.3.5.2 Appointing Maintenance Strategy Based on Failure Modes

Once the failure modes, effects, and root causes have been defined and the RPN determined for the major equipment failure modes, we can determine the proper maintenance strategy for the equipment. As already written, there is a new approach used that considers the failure modes separately to determine the maintenance strategy, and not the asset as whole with its failure modes. The approach is based on separate consideration of all failure modes of an asset. It uses the information gathered with previously conducted FMECA. A decision route has been developed [32, 34] to go through for all failure modes and to determine which of the existing maintenance strategies would be optimal as actually the **ultimate goal of maintenance is to prevent any of the major failure modes to happen!**. A decision **route was developed based on** [34] Maintenance decision diagram, incorporating proactive maintenance also, **as a logical tree helping in deciding** the best applicable maintenance strategy based on the analysis of the major failures. The principle is shown in Fig. 2.5. When carrying out the decision route for all failure modes separately, we have the benefits of performing an in-depth and more comprehensive

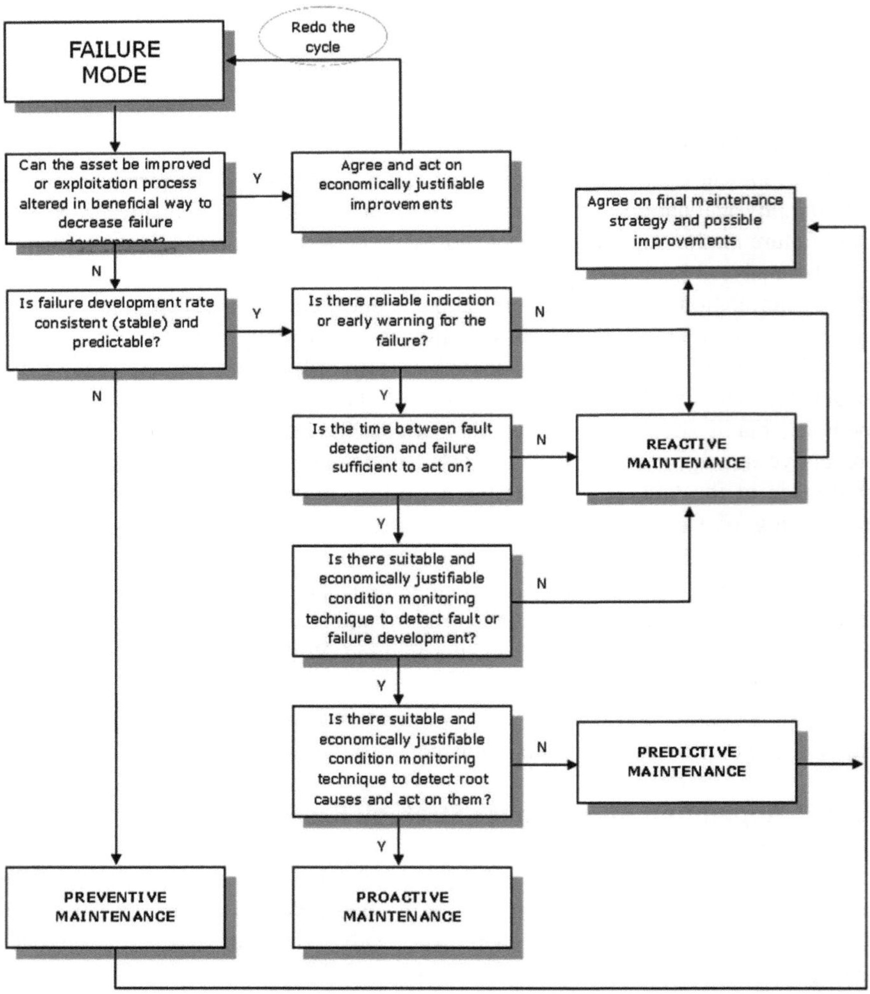

Fig. 2.5 General decision diagram on maintenance strategy

analysis of the failure mode than at the multiple bases, especially to focus on practical issues as available condition monitoring techniques, their reliability etc.

Steps to go when deciding on the maintenance strategy are as follow:

Determine the deterioration rate. If it is fixed and predictable and there is no need for design modifications, than a preventive maintenance is best solution.

If this is not the case, but if there are early reliable indication with sufficient time interval for action, there is available monitoring system and there is possibility to act on root causes, then proactive maintenance is the best strategy.

If we cannot act on root causes, then predictive maintenance is a more suitable strategy. If there are indications for failure, but the maintenance is not cost-effective, than still the old "run-to-failure" is best applicable strategy. At the end, when considering the equipment, the maintenance actions should be a set of actions handling each failure at the estimated way. Many of the maintenance activities are common for more maintenance strategies, but the estimation of the findings is different and actions undertaken afterwards are different.

Following this basic principle, examples have been worked out with application of decision route for the major faults of cement rotary kiln determined with FMECA. The decision routes for the bearing boundary lubrication and kiln crank as one of the major kiln faults are shown on Fig. 2.6a, b.

For most of the major failure modes at cement rotary kiln, proactive maintenance strategy is feasible. In some cases it is easy and available, but in some cases as crank and boundary lubrication; proactive maintenance would be possible if proper measuring method/technique could be applied. It is not always the rule to drive for proactive maintenance, but if all other facts are implying the necessity for that strategy, except some obstacle, it should be considered seriously.

Considering the effects of the crank as failure mode (FM) and its RPN, it is worth to investigate more on what is required to do to enable proactive maintenance strategy. The first issue to analyze is about the predictability of the time-to-failure. It was already given that this time is not consistent and sometimes it can be as long as few weeks, but sometimes it can be as short as few hours. We cannot predict and improve it on fix-time base. Possibility for a reliable indication or early warning is next that should be determined. In this example, the kiln condition changes, as a consequence from kiln crank state in a way that it changes the supporting roller deflections by reflecting the load cyclic change due to crank action (see Chap. 7). If this deflection change is detected early enough, then there is enough time to act on the failure development. The next, very important aspect is to see the practically the feasibility of the monitoring system. It is about being able to measure the indicators, or finding **suitable and cost effective condition monitoring technique to be employed**. This is of crucial importance. With today's widely used typical, conventional monitoring methods at kilns there **is no suitable** system that would give us early enough information to act on the failure development. Such state could bring us to an "operate to failure" strategy, which as already stated can cost the plant more than hundred thousand euros. It is exactly the proactive condition monitoring approach that is also subject of this book that makes it possible to

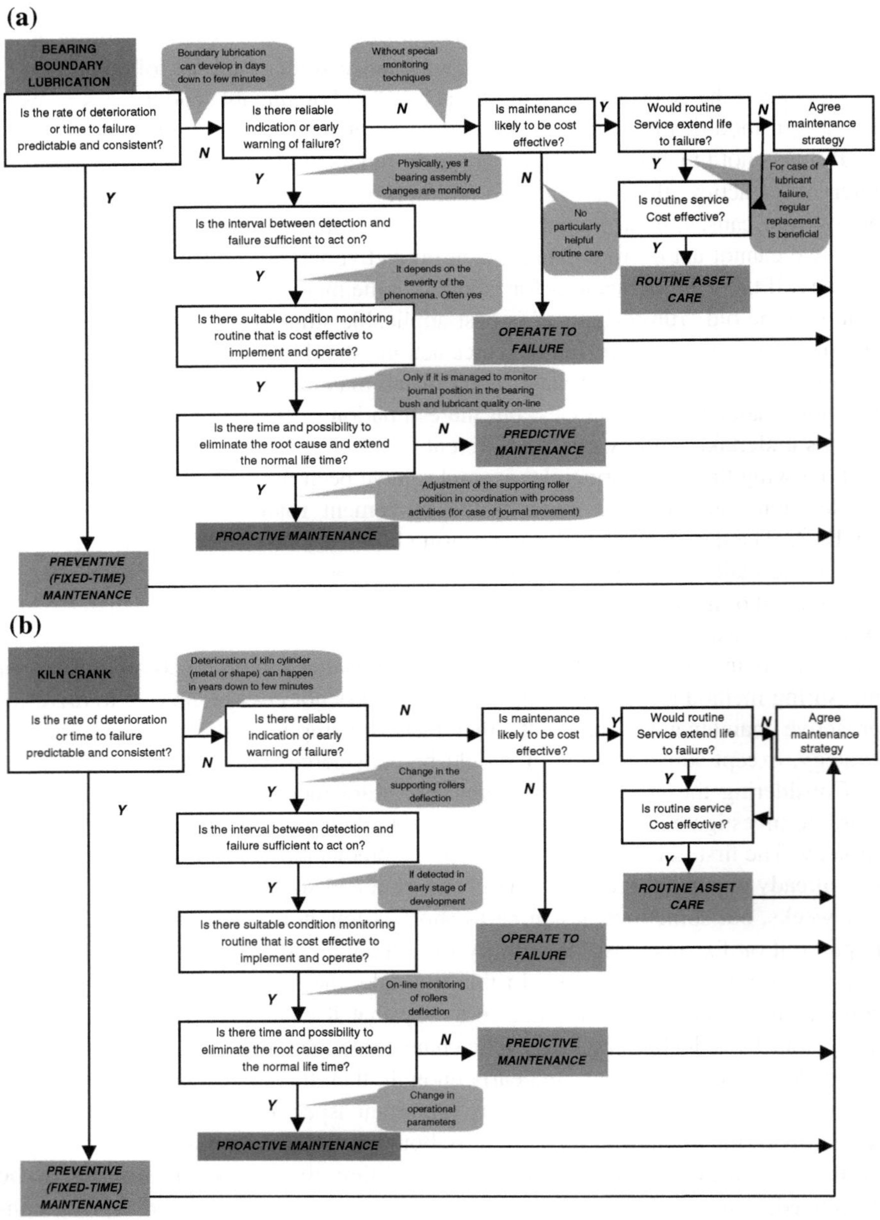

Fig. 2.6 **a** Example of decision diagram for maintenance strategy for bearing boundary lubrication failure mode. **b** Example of decision diagram for maintenance strategy for kiln crank

monitor the indicating fault root cause in real-time described as example in Chap. 7. Once providing such online information, there are possibilities to eliminate the root cause and prolong the lifetime of the equipment, which will be discussed later.

References

1. A. Crespo Marquez, P. Moreu de Leon, J.F. Gomez Fernandez, C. Parra Marquez, V. Gonzalez, The framework: a practical view to maintenance management, in *Safety, Reliability and Risk Analysis: Theory, Methods and Application*, ed. by Martorell et.al. (Taylor & Francis Group, London, 2009). ISBN978-0-415-48513-5
2. T. Davcev, Nadeznost I odrzuvanje na tehnickite sistemi, NIP "Studentski zbor" (Skopje, 2009)
3. J. Moubray, *Reliability Centered Maintenance*, 2nd edn. (Industrial Press Inc., New York, 1997)
4. E. Deloux, B. Castainer, C. Bérenguer, Condition-based maintenance approaches for deteriorating system influenced by environmental conditions, in *Safety, Reliability and Risk Analysis: Theory, Methods and Application*, ed. by Martorell et.al. (Taylor & Francis Group, London, 2009). ISBN978-0-415-48513-5
5. J. Luyk, J.L. Rouvroye, Further validation of organizational influencing factors of proactive risk management, in *Reliability and Risk Analysis and Safety* (Taylor & Francis Group, London, 2010)
6. S. Gassner, Deriving maintenance strategies for cooperative alliances—a value chain approach, Germany, Internet source
7. R.T. Buscarello, *Practical Solutions to Machinery and Maintenance Vibration Problems*, 4th edn. (Update International, Inc., Lakewood, 2002)
8. L.M. Bartlett, E.E. Hurdle, E.M. Kelly, Comparison of digraph and fault tree based approaches for system fault diagnostics, in *Safety and Reliability for Managing Risk*, ed. by C. Guedes Soares, E. Zio (Taylor & Franics Group, London, 2006). ISBN 0-415-41620-5
9. G. Birkeland, S. Eisinger, T. Aven, Risk based maintenance prioritisation, in *Safety, Reliability and Risk Analysis: Theory, Methods and Application*, ed. by Martorell et.al. (Taylor & Francis Group, London, 2009). ISBN978-0-415-48513-5
10. Preventive Maintenance/Essential Care and Condition Monitoring, IDCON, Inc, Edition IV, 2003
11. R.P. Chapman, *Recommended Procedures for Mechanical Analysis of Rotary Kilns* (Fuller Company, Bethlehem, 1985)
12. G. Pantazopoulos, Operational and strategic planning for continuous improvement of RCC Technical Department, Roanoke 2002
13. J. Johansson, M. Rudberg, *Maintenance Management in Process Industries: Mapping Current Practices*. (Department of Management and Engineering, Linköping University, Sweden)
14. M. Bengtsson, On condition based maintenance and its implementation in industrial settings. Malardalen University Sweden, Press Dissertation No. 48, 2007
15. I.T. Castro, K.T. Huynh, A. Barros, C. Berenguer, Maintenance strategies in dynamic environments for reairable systems with minimal repairs dependent on their deterioration level, in *Reliability and Risk Analysis and Safety* (Taylor & Francis Group, London, 2010)
16. Richard M. Barrett, *Low Frequency Machinery Monitoring Measurement Considerations*. (Wilcoxon Research, Inc., Germantown)
17. J. Mais, *Spectrum Analysis: The Key Features of Analyzing Spectra* (SKF Reliability Systems, USA, 2002)
18. J. Mais, *Low Speed Analysis: Vibration Monitoring of Low-Rotational Speed Applications Using Accelerometers* (SKF Reliability Systems, USA, 2003)

19. SPM: ILearnVibration, http://www.ILearnInteractive.com
20. M. Samrout, E. Chatelet, R. Kouta, N. Chebbo, Optimization of maintenance policy using proportional hazard model, in *Safety and Reliability for Managing Risk*, ed. by C. Guedes Soares, E. Zio (Taylor & Franics Group, London, 2006). ISBN 0-415-41620-5
21. S. Wichtendahl, R. Denton, *Cost Effective Solutions*. (Wilcoxon Research, Inc., Germantown)
22. IEC60812 FMEA
23. E.A. Colosimo, F. Pontel, RCM program evaluation in a cement company, in *Risk, Reliability and Society Safety*, ed. by T. Aven, J.E. Vinnem (Taylor & Franics Group, London, 2007). ISBN 978-0-415-44786-7
24. ISO13379 Condition monitoring and diagnostics of machines-General guidelines on data interpretation and diagnostics techniques
25. Lifetime Reliability Solutions: Maintenance and Asset Management Strategy Using Physics of Failure Factors Analysis, Internet source
26. J. Roberts, Total Productive Maintenance (TPM), Technology Interface/Fall 1997 USA
27. J. Wardhaugh, Extract from 2004 Singapore IQPC reliability and maintenance congress presentation 'maintenance the best practices'—www.lifetime-reliability.com Useful_Key_Performance_Indicators_for_Maintenance
28. D. Hutchins, What is total productive maintenance? David Hutchins International Quality College
29. G.S. Norat, BSEE, PE Senior Engineering Consultant PdMtech, Inc. 09 Nov 2008—A PdMtech Inc. White Paper
30. A.A. Andreani, E. Zio, F.K. Rodriguez, P. Viveros, Integrated analysis of system reliability and productive capacity of a production line, in *Reliability and Risk Analysis and Safety* (Taylor & Francis Group, London, 2010)
31. Polysius: Technology Forum, Neu Beckum, 2004
32. Z. Stamboliska, Proactive method of low-speed machines condition monitoring. Ph.D. thesis, Wroclaw University of Technology, Poland, 2011
33. FLSmidth Institute: The International Maintenance Seminar, Copenhagen, vol. I, 2005
34. ABB Eutech Assets

Condition Monitoring Considerations for Low-Speed Machines

3

Abstract

Condition monitoring of low-speed machines differs from medium to high-speed machines mainly due to misunderstanding of the condition changes and due to low energy emitted when certain damages happen in the moving components. Failures develop slowly and they stay latent till some critical point of their development interval when it is usually too late to act on prevention. Therefore, special considerations must be taken when deciding on data collection and special measuring systems employed for condition monitoring of machines with low cycles of operation. These considerations are presented in the chapter in order to assist technical teams to successfully implement effective maintenance systems to such machines.

Keywords

Low-speed machines · Condition monitoring · Impact energy · Nonstationary conditions · Data collection · Measuring systems · Probes · Sampling rate · Filtering · Frequency · Amplitude · Vibrations

3.1 Key Differences in Condition Monitoring of Low-Speed Machines

There are a number of applications in today's industrial and manufacturing environments for machines that operate at low-speed. Although there are no universally-accepted criteria for classifying machinery as low-speed, intermediate speed, or high speed, it is generally recognized that low-speed machines are those that operate at or below 600 rpm. These applications include paper machines, cooling tower fans, wind turbines, rotary dryers, low-speed agitators, ball mills, crushers, rotary kilns, excavators, some process fans, and mining machinery. Some high-speed machines can also exhibit faults in the region of low-speed such as gearboxes, compressors, and others, and therefore considerations of low-speed machinery are also valid here.

Z. Stamboliska et al., *Proactive Condition Monitoring of Low-Speed Machines*,
DOI 10.1007/978-3-319-10494-2_3

The efficient operation and maintenance of these machines is essential to maximize the production and minimize the downtime. However, latent or incipient failures of these machines often go undetected due to two main reasons:

- Frequently present people's mindset related to low-speed machinery "saying" that nothing can happen so fast as not to be detected even only by subjective monitoring techniques as simple as visual inspection or listening; and
- Usage of inadequate monitoring techniques and measuring instruments that are appropriate for high-speed machines, and not sensitive enough to low frequency changes.

Actually, at the low-speed machinery changes happen slowly so that if monitoring left only to subjective monitoring or widely accepted CM techniques, people could get used to these small incremental changes "forgetting" how the state was when machine was totally free of any faults.

The advantage of the low-speed machinery is that faults usually do not develop so fast into catastrophic failures, but it is also their main weak point because many "signs" go undetected and accepted to be a normal state. Here, the P-F interval (development of potential failure to functional one) could be very long, but the early stages of progress are not so evident.

This leads to situations as cracks in housings, shells, trunnions, divot, or flat spot in low-speed bearing appear and it seems as "sudden" event, while the pointing signs were all the time there, at the machine.

Condition monitoring of slow speed machines reflects to the same actions as regular condition monitoring, or data collection, data processing and analysis, interpretation, and finally diagnostic of the machine condition. Why then special considerations for the low-speed machines? There are two main aspects that should be considered:

- **Low speeds create lower overallimpact energy** generated by various irregularities during relative moving components at low speed; and
- **In most cases, low-speed machines** are part of heavy duty industries and conditions of exploitation are mostly **nonstationary** thus making the traditional CM approaches not always appropriate.

Reflecting these basic differences into the phases of CM, it requires special attention to be paid on all phases.

The lower energy generated when faults exist means that it is more difficult to recognize any changes in the machine's condition, especially if CM system is tuned as for high-speed machines. Lower energy affects all phases of CM techniques related to movement of the machines components. At first, the sensing of the changes is more difficult not only because the lower energy creates less visible amplitudes and peaks in the recorded signals, but also it can be easily 'blended' into

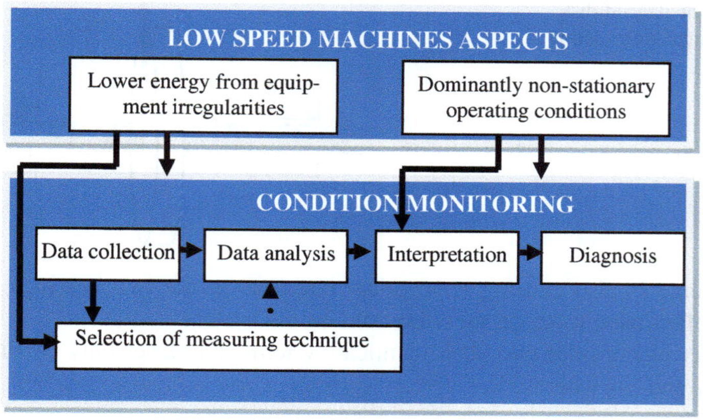

Fig. 3.1 Low-speed machines properties affecting CM phases

the noise from the environment, thus losing all important parameters. Figure 3.1 illustrates what aspect of low-speed machine affect what segment of CM system.

Lower energy makes more difficult to collect the right data, so this is to be considered not only in general when planning the CM system, but also more specially when deciding on data collection technique. This is especially valid for vibrodiagnostics where the impact energy is the primary source of signals, so the probes have to be sensitive and with such frequency range that will include the cycles of interest (this will be more elaborated later in this chapter). Once we have the data, they should be filtered out from any unwanted components and noise and it is of extreme importance to apply appropriate data analysis and interpretation. Selecting wrong measuring technique or even a probe will give false data for analysis, so the final interpretation and decision of the machine health will be wrong.

Dominant nonstationary conditions do not affect much the selection of measuring technique, but affect the overall organization of the measurements (when to be made, how to relate them to the real operational speeds and loads), and at most, the interpretation of the results must consider the operating conditions, possible peak loads, or turbulences that do not necessarily mean discrepancies in the machine condition.

To illustrate at more tangible way the meaning of the lower energy at these machines compared to high-speed machines, let us consider one example of an imbalanced disk (with mass m) on a shaft rotating with constant angular speed ω, supported on two bearings with its center of gravity located at certain radial distance e (eccentricity) from the centerline of the shaft (Fig. 3.2). The effect of this imbalance is centrifugal force acting on the disk:

$$\mathrm{Fc} = m\omega^2 e \tag{3.1}$$

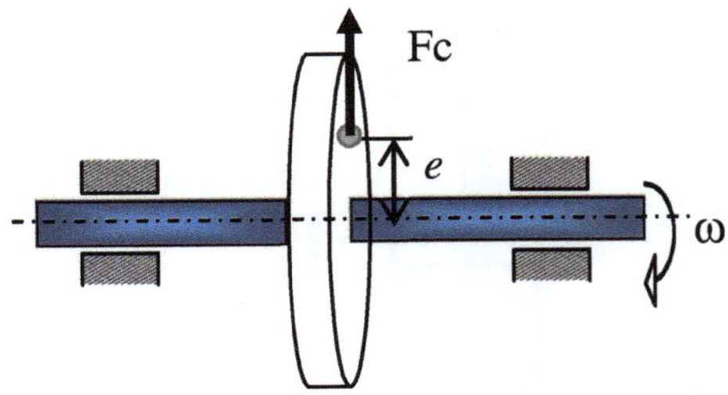

Fig. 3.2 Imbalanced disk and centrifugal force acting

This force is also causing appropriate reactions in the two bearings, which is actually often used to sense the vibrations in the system.

Now let us consider two such identical systems that differ only by the angular speed: ω_1 and ω_2 where:

$$\omega_1 = 1{,}000\,\text{rpm}$$
$$\omega_2 = 100\,\text{rpm}$$

$$\text{Correspodingly}\quad \frac{\text{Fc}_1}{\text{Fc}_2} = \frac{m\omega_1^2 e}{m\omega_2^2 e} = \frac{m10^6 e}{m10^4 e} = 10^2$$

Force acting at the slower disk would be 100 times less than the high-speed disk, although the speed difference is only 10 times!

Similarly, it happens when there is impact of damaged bearing component during rotation of the shaft. The impact energy at any repetitive motion of damaged component when moving along another component that is also creating vibrations is as follows:

$$E_k = 1/2\,mv^2 \tag{3.2}$$

where

- m = mass of the component (kg)
- v = velocity of the component (m/s)

The impact energy deviates with the square of the velocity at the same mass of the components.

These are only two examples showing that energy generated at slow speed machines due to various irregularities of the moving components is significantly lower than at high-speed machines, thus making more difficult faults to be recognized. It is, therefore, evinced why careful selection of measuring technique and probes, as well as signal processing should be made as elaborated in Sect. 3.2.

The other aspect that was mentioned was the nonstationary conditions. Although this is not valid only for low-speed machines, it is gaining major importance at these machines because in most cases slow machines are employed in harsh environment of quarrying or milling processes. Wind turbines, rotary excavators, crushers, and other mining and civil engineering machines are just some of the examples where loads can vary significantly during exploitation. Actually, researchers and applicants have become very much aware and recently there are new trends of specialized examinations on the condition monitoring in non-stationary conditions. At the specialized conference held in 2013 [1–3], the analysis made on the present state of CM clearly points that most established machine condition monitoring methods are related to stationary operations. But many machines work in nonstationary operations and they are characterized by varying rotation speed and load, and thus vibration generated are load dependent, which gives load dependent diagnostic features when using: spectrum and cepstrum analysis, order tracking, cyclostationary analysis (Spectral Correlation), short time Fourier transform, Wigner-Ville distribution, crest factor etc. For proper machine condition monitoring there is a need to identify instantaneous frequency and load. In cases when the relationship between the rotational speed and load is known, the fault feature estimation may be used. Also a need was recognized to develop ways of signal analysis, which will take into consideration that machine is a unity of many different elements influencing each other (bearings, gears, and shafts) and degradation of only one specific element or fault is not enough. Use of mathematical modeling and simulations, together with experimental investigations etc. to create the so-called pattern recognition for condition monitoring of certain components working under varying load has become new and useful trend in condition monitoring.

3.2 Aspects of Condition Monitoring Phases for Low-Speed Machines

3.2.1 Data Collection

The important note is that when dealing with low-speed machinery, it is of primary importance to understand the physical nature of the machinery components function, potential failure causes, and the best describing monitoring parameters. Otherwise, very wrong results could be obtained misleading the maintenance team and also concealing the development of potential failures.

The condition monitoring has many techniques (as described in Chap. 5), and all of them start with data collection. But it is clear that low speed makes differences at those techniques related to the effects of the moving components. Therefore, for the monitoring of dynamical effects, specially the vibrodiagnostics, where knowledge is required, special attention should be paid on the phase of data collection and selection of measuring instruments. So, further attention is focused on the considerations for vibrodiagnostics.

The value of 600 rpm already mentioned, comes from its representation of the approximate speed at which alert levels relative to vibration in the velocity domain must be reduced as the speed of the machine decreases.

A commonly accepted methodology is to establish alert levels, which decrease linearly with decreasing speed (or constant in the displacement domain).

Low-speed machinery, or low-frequency condition monitoring normally requires measurements within 0.1–10 Hz (6–600 cycles per min) bandwidth, or often even lower than 0.1 Hz. Actually, motion below 10 Hz produces very little vibration in terms of velocity, and very large vibrations in terms of displacement (Fig. 3.3).

The most important issues to be considered at low-speed machinery is the measuring system employed, or in particular:

- measuring probes
- sampling rate
- filtering data.

3.2.1.1 Measuring Probes

There is no clear picture about measuring probes to be used when measuring vibrations, specially for low-speed machines. There are many different types and manufacturers with various advantages and disadvantages. In general, two main type of probes can be used:

- Accelerometers;
- Displacement probes.

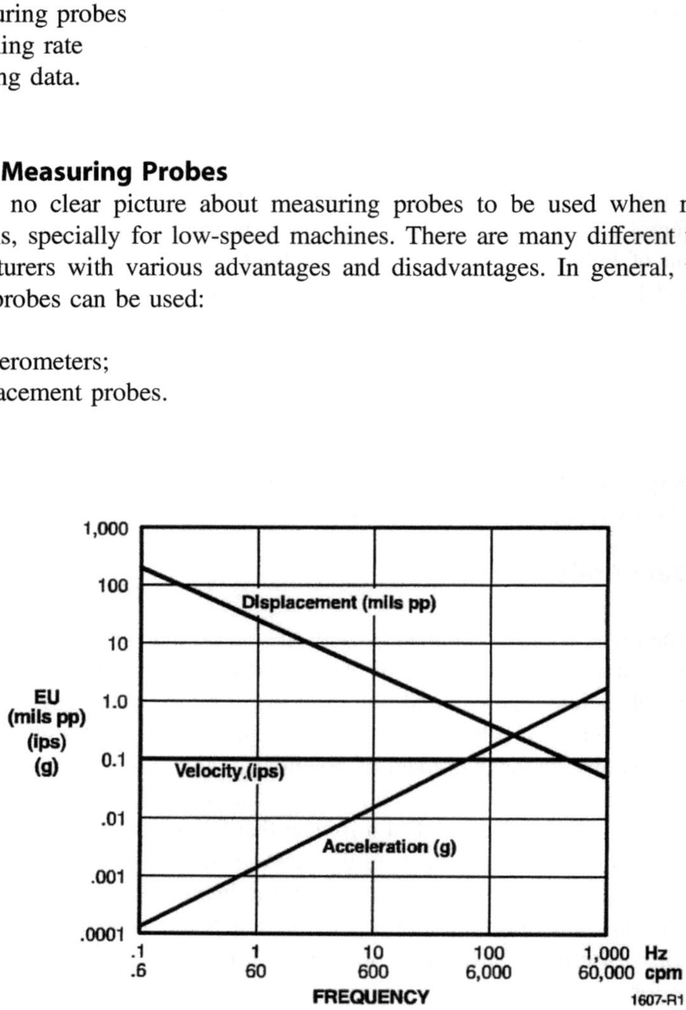

Fig. 3.3 Selection criteria for monitoring dynamical effects of different range of speeds (frequencies)

Accelerometers

They are tools for acceleration measurements. There are many different types of accelerometers for industrial and scientific applications. This can give the advantage for engineers to use the wide range of accelerometers for different applications such as condition monitoring and short term tests. One of the most important classifications of accelerometers is based on the signal response, which can be:

- AC type—cannot measure static acceleration, but they are appropriate for dynamic tests. They are also useful for calculation of velocity and position of the tested equipment by integrating the acceleration curve. The most popular AC accelerometers are equipped with piezoelectric elements as their sensing mechanism.
- DC type—can sense constant acceleration such as gravity. Therefore, they can measure vibrations down to 0 Hz theoretically. The most popular DC accelerometers are capacitive MEMS and piezoresistive sensors types.

Another classification related to technique used for acceleration sensing is as follows.

- Piezoelectric—piezoelectric accelerometers use piezoelectric effect. When piezoelectric materials are deformed by acceleration forces an electric charge is generated on the piezoelectric materials. This charge is proportional to the applied acceleration. In such accelerometers, piezoelectric material is used as an active element. One side of the piezoelectric material is connected to rigid base, while seismic mass is attached to the other side. The produced charge is usually converted to voltage using integral electronics—charge amplifiers and associated signal conditioning circuit. These accelerometers are AC response sensors type. They are characterized by wide dynamic range and frequency bandwidth.
- Piezoresistive—piezoresistive accelerometers use piezoresistive effect, which is related to the resistivity change of a semiconductor due to applied mechanical stress. These accelerometer uses a piezoresistive substrate in place of the piezoelectric crystal. The applied force caused by the seismic mass, changes the resistance of the crystal, which is monitored through the whetstone bridge used. This resistance change is proportional to acceleration. Piezoresistive accelerometers are DC response sensors type. They have true DC response. They can be designed to measure up to ±1,000 g. They are preferred in high shock applications.
- Inductive—in these accelerometers, proof mass is made of ferromagnetic materials. The proof mass is designed in the form of core, which can move in or out of the coil. The coil impedance is a function of the applied acceleration. Inductive accelerometers are DC response sensors type.
- Capacitive—capacitive accelerometers sense a change of electrical capacitance, with respect to acceleration. When acceleration is applied to the sensor, the proof mass displaces from its nominal position, causing an imbalance in the capacitive

half-bridge, which is measured using charge integration in response to a voltage pulse applied to the capacitor. Capacitive effect is most commonly used in MEMS accelerometers. Capacitive accelerometers have true DC response, but limited frequency range and limited dynamic range, which are not important in low frequency testing applications. They are actually recommended for low amplitude and low frequency events monitoring. Signal to noise performance is generally very good in this type of accelerometers. The bandwidth of capacitive accelerometers is usually limited to a few hundreds of Hz.

The most important parameters of accelerometers, which can help in choosing proper equipment for condition monitoring, are:

- Sensitivity—sensitivity is the ratio of change in acceleration (input) to change in the output signal. It describes the gain of the sensor and can be determined by applying 1 g acceleration (by rotating the sensor for 180°) and recording the output value before and after rotation. Dividing the output by 2 gives the sensitivity of the sensor. Sensitivity is specified at a particular supply voltage and is typically expressed in units of mV/g.
- Measurement range—acceleration range is usually given in units of gravity constant. 1 g is equal to the gravity at sea level.
- Frequency range—the value describing applicability of sensor in the unit of Hz.
- Broadband resolution—generally, resolution is the smallest amount of acceleration, which can be indicated by sensor. It is important to know that resolution is not accuracy.
- Output signal—there are different outputs such as voltage, current output, charge output, and IEPE. There is also general difference in output type such as analog and digital.
- Number of axis—there single, double, and triaxial accelerometers. The three axis accelerometer will measure acceleration in three orthogonal directions.

Condition monitoring of low-speed machines requires measurements of low frequency motion. In such a case MEMS accelerometers are usually required. Unlike piezoelectric accelerometers, these sensors respond to very low, close to 0 Hz frequency. The typical sensitivity for low frequency MEMS accelerometers starts from 1,000 mV/g.

However, in case of very low frequency motions it is necessary to use displacement measurements technique to enable proper quality of condition monitoring.

Displacement Sensors

Similarly to accelerometers there are many different types of displacement sensors available for industrial and scientific applications. One of important classification of displacement sensors is based on the principle of operation:

- Contact sensors—operation of such sensor is based on the displacement change sensing by the direct contact of sensor with measured object. They are prone and sensitive to wear and quality of the measured surface.
- Noncontact sensors—there is no direct contact of the sensor with measured object. Most popular sensors use electrical or magnetic field change phenomena to sense displacement change. They are wear-free sensors.

In the condition monitoring applications most of the sensors are noncontact sensors. The most popular noncontact sensors for displacements measurements are listed below:

- Capacitive sensors—measure displacements, distances, or positions of any electrically-conducting objects. They operate based on capacitance electrical property, which is created by applying an electrical charge to two conductive separated objects. Capacitive sensors are available for a wide range of applications. However, they require good operational (clean) condition. They enable high resolution measurements, and have good temperature as well as long-term stability. Typical measuring range is 0.05–10 mm.
- Magnetic-inductive sensors—the most popular inductive sensors are Eddy-Current sensors, which operate with magnetic field and are capable to measure position or change of position of any conductive objects. Position change creates an alternating current in the sensing coil of the probe. The alternating magnetic field then induces small currents (eddy currents) in the conductive material. The interaction of the magnetic fields represents the distance change between probe and the object. As the distance changes, the sensor sense the change in the field interaction and generates a voltage output proportional to the distance. Eddy-Current sensors have little lower measurement resolution compared to capacitive sensors. However, the advantage of their usage is that they can operate in the dirty condition. Typical measuring range is 0.5–80 mm.
- Laser focus type sensors—they operate according to the triangulation principle. A light spot is projected onto a measurement surface with the use of a laser. By means of receiving optics, the reflection is mapped onto a light-sensitive element. The distance to the object can be determined, based on the position of the mapped light spot. Laser sensors measure displacement, dimension, distance, and position of any objects (metallic, nonmetallic, conductive, nonconductive, etc.). They have wide measuring range, which starts from 2 mm and can go up to 1,000 mm. They require good operational (clean) condition and enable high resolution measurements.
- Laser distance sensors—they are appropriate for long distance medium accuracy measurements up to 250 m distance. They operate on the time of flight principle by sending a laser pulse toward the object and measuring return time. This technique is not appropriate for high precision measurements.
- Confocal sensors—they use white light confocal displacement sensor and operates on optical principle of measuring the reflected light's component wavelengths. They enable measurements of any object also shiny or transparent

surfaces. They have very good accuracy. They operate around measuring center distance, which can reach 50 mm approximately. The typical measuring range is not higher than 20 mm. They require clean operational conditions.
- Ultrasonic wave sensors—they transmit a short burst of ultrasonic sound toward a measured object. When the sound is reflected, it returns to the sensor as an echo. The distance between the ultrasonic linear position sensor and the target is calculated from the signal's return time and the propagation velocity of the measurement medium. These sensors require a minimum distance to provide a time delay so that the "echoes" can be interpreted. Typical measuring range is 0.1 m up to few meters. They have low accuracy and require large detecting point area.

Among all presented noncontact displacement sensors, Eddy-current sensors are the most popular and appropriate for condition monitoring of low-speed machines. This is due to the fact of their high accuracy, good response speed, high durability, and possibility of usage in dirty condition. They have also one very important advantage, which is acceptable price of the sensors.

Having in mind all those types of probes developed with various properties, when deciding the monitoring system, the main focus should be paid to the following ambiguities:

(a) *Reduced sensitivity at low frequencies*. The sensitivity of standard accelerometer/instrument combinations used today is significantly reduced at frequencies below the 120–300 cpm range (2–5 Hz). Actually, the most commonly used piezoelectric accelerometers that rely on continuous movement to excite the sensing crystal have low frequency operational detection limit of about 2 Hz (120 cpm) before the signal decay. Very good explanation was given by Mais in [4] who explains that this is caused by electronic noise, which becomes a significant component of the total low frequency signal as the very small acceleration signals diminish with frequency.

The higher amplitude, higher frequency components of the vibration signal will tend to drown out the low amplitude, low frequency vibrations of interest, producing an unacceptably high signal-to-noise ratio. Further complicating the processing of vibration signals produced by low-speed machines is that the sensor (accelerometer) adds an electrical noise component to the composite signal. The problem for the monitoring system [5, 6] becomes one of extracting the low level components (requires a wide dynamic range) from the composite signal and then differentiating the vibration components of interest from the other low level components generated from electrical noise sources, temperature transients, and other similar sources of irrelevant components. Thus, the analysis of the vibration signal must become more sensitive to the low level (and often low frequency) components when monitoring low-speed machinery. According to [4, 7] typical analysis techniques for medium and high-speed machinery are not able to effectively adjust to these higher sensitivity requirements and process the vibration signal without eliminating the low level components. Many instruments use filters to minimize the noise, but this can mask the true signals as well. Also actual working range will also be affected by operating conditions, such as mounting method, temperature, and stabilization time. Special accelerometers and instruments are capable of measuring down to 6–10 cpm (approximately 0.1 Hz), but these are usually reserved for special applications due to their expense. Still, in the last few years researches have been conducted and today there is availability of also less expensive instruments for monitoring damages at low-speed bearings, gear, and similar components even down to 0 rpm.

For example, the novelties from the shock pulse measurements with the so-called "High Definition" method [8] are application of advanced digital algorithms that provide very high dynamics, enabling distinguishing the desired signal from background noise. Signals are picked up and enhanced, resulting in clear and unobstructed information for the machine condition. These systems are already applied successfully in applications, such as twin wire presses, screw presses, drying cylinders, and wind turbines, some running at only a few revolutions per minute.

Another very important aspect is that measuring with handheld equipment is not advisable for low RPM applications. Actually, at low speeds, damages appear and develop in phases. At the very beginning, there may be spalling or other crispy edges damages that produce higher energy peaks that are easier to be recognized as shock pulses. But, with the time, due to slow movement, these edges are worn out and turn into new transitional form that give no more such spiky energy. So, if measured with not continual method in exactly such period, damages may be omitted.

(b) *Instrument integration errors*. As previously mentioned, it is usually recommended displacement units to be used for measuring low frequency vibration. However, a potential problem arises when very low frequencies are received by accelerometers and converted from velocity through single integration, or displacement through double integration. Integration can cause errors on the vibration spectrum. Following some researches [4, 7, 9], these errors are evident as large "spikes" that occur at the lowest frequencies usually within the first three lines of resolution (sometimes called "spectral bins").

The sensitivity required of the sensor is based on the amount of displacement to be sensed. For high sensitivity applications, the sensor of choice for vibration analysis is typically an accelerometer having a dynamic range of between 100 and 120 dB. It is desirable that the analysis method be able to accommodate this range.

Most analyzing methods, as Klubnik [9] and Mais [4] and others [10–12] wrote, result in an attenuation of the accelerometer signal so that the dynamic range of the analyzed signal is, for example, reduced to about 80 dB. Thus, there is a need for a measurement system having a dynamic response that is at least as good as that of the accelerometer.

It means there is a need for a reliable and effective measurement system for monitoring low-speed machinery that is capable of differentiating fault-induced vibration components from machinery-induced vibration components as well as other extraneous components, including electrical noise and transients. The system should be minimally responsive to temperature transients and maintain a wide dynamic range.

Depending on the application, nowadays to monitor low-speed machinery special displacement sensors are used or accelerometers with low-frequency range or the shock pulse transducer that is a piezo ceramic transducer with resonance frequency as high as 32 kHz.

3.2.1.2 Sampling Rate

The basic pre and post processing operations shall be properly executed to enable further, often complex operations on collected measuring data, obtained from measuring system (accelerometers or displacement sensors, recorder, analyzers etc.). In signal processing, sampling is transformation of continuous (analog) signal to a discrete signal. The number of samples taken in one second of the signal is called sampling frequency f_s and it is expressed in Hz. The required sampling frequency is strictly related to the frequency of the signal we monitor. This rate should be set in such a way so as to enable reconstruction of analog signal from discrete data without losing information on the measured parameter. Theoretically, in order to make such conversion without loss of information, there must be the Shannon-Kotielnikov condition fulfilled. It says that the sampling rate cannot be less than twice of the highest frequency (Nyquist frequency) of measured analog signal. If this condition is not fulfilled, then the phenomenon of aliasing is present. However, for condition monitoring the minimum sampling rate must be much higher than the one given from Shannon-Kotielnikov equation. Such procedure is called oversampling. The proper minimum sampling frequency in such case can be estimated based on test on the measuring object. Several tests with different sampling rate have to be performed and comparison of converted signals gives information on the loss of information in relation to sampling rate. Based on such tests we can get minimum sampling rate. In the stationary signals obtained from low speed condition monitoring systems this frequency is 5–10 times higher than Nyquist frequency, which in the low-speed machinery it is 10 Hz (600 rpm).

3.2.1.3 Signal Filtering

Operation of low-speed machinery requires many types of equipment, which creates, vibrations, noise, and similar effects. Such equipment can be, e.g., electrical motors, fans, and pumps. A wide band noise is created by this equipment and disturbs visual interpretation of obtained signal and further analyses as well. Therefore, filtering of measured data is used to solve these problems. For the low-speed machinery the most commonly used filter is low pass filter. Its enables to cut higher, not useful, frequencies, which come from operation of the surrounding equipment. Depending on the measuring system, filtering can be done in the post processing off-line operations or online during tests.

The most important setting in the low pass filter is its frequency f_{lp}. If it is set too high the signal is affected by noise of higher frequencies. If set too low, important data can be lost. Theoretically, this frequency is related to the Nyquist frequency f_N:

$$f_{lp} \leq f_N$$

Considering efficiency of filters available in the monitoring systems, this frequency is usually lower and equals:

$$f_{lp} \leq \sim 0.5 f_s = f_{Np}$$

where f_s—sampling rate, f_{Np}—practical Nyquist frequency.

The final low pas filter frequency can be obtained based on the several tests of the same signal with different frequencies settings.

3.2.2 Data Analysis and Interpretation of Results

Same approaches are being used as with medium and high-speed machines, and thus values obtained expressed through velocity (mm/s) or acceleration (mm/s^2) are of low value as they "cover" the faults developing in the machines. But this is also not totally true as these techniques are justified in machines where a part from main components running with low-speed, there are also components (such as drive trains consisting of pinion and girth gear) exhibiting higher vibration frequencies (like gear meshing frequency) that are suitable to be determined with velocity/ acceleration measurements or analyzing techniques.

Some of the changes in low-speed machines, like wearing, flat spots, misalignment, orbiting, etc. can be somehow monitored by visual inspection, but of course more accurate and really representative data would be gained by certain geometrical or other dynamical measurements in real time. Here, exactly we face the main problems: **how to detect accurately so slow changes and how to link them with the real causes**?

At medium and high-speed machines, vibration analysis is almost fully in-charge and gives all required data for early detection of faults. Actually, there are many known techniques [10, 13–18] that have been developed on a very high level for monitoring medium and high-speed machinery by analyzing mechanical vibrations generated during operation. An accelerometer is typically used to generate a signal corresponding to the vibrations produced. The signal produced by the accelerometer will include normal (nonfault) vibration components as well as fault related vibration components occurring within the machine. Various analytic techniques are then used to separate the vibration signal into its constituent components and identify those components corresponding to possible faults and other mechanical vibrations.

Unfortunately, the techniques, which are used to monitor faults at medium and high-speed machinery, are not equally effective for monitoring faults within low-speed machinery. There is a special difficulty in monitoring low rpm vibration because the low frequency components will typically be of lesser amplitude than the higher frequency components. Signals originating from low-speed bearings, gears, and other components are typically low in energy content, disorderly, and therefore it is very difficult to analyze them. Actually, these signals are often mistakenly dismissed as being background noise and often real damages go undetected until real failure happens.

Actually, a different approach to vibration limits and evaluations is needed for low-speed machines. The lower the speed or frequency being measured, the more precautions are needed. Following many sources [4, 9, 10, 13, 19–21] and also personal experience of the authors of this book, for low-speed machines, displacement is the preferred measuring unit. For example, if there is a machine with 1,800 rpm of the drive motor and vibration velocity measured was about 5 mm/s and the reading measured on a 3,600 rpm motor was also 5 mm/s, then both machines would be considered to have the same vibration intensity. Theoretically, both machines were receiving the same amount of "punishment" as shown by their vibration amplitudes. If speeds were lower, such as 1,200 rpm, or higher such as 10,000 rpm, the same evaluations would be made based on the velocity amplitudes. However, as machine speeds get lower and lower, a problem does occur which, if not properly dealt with, can produce seriously defective judgments regarding the machine's condition as shown by vibration readings.

In the theory of vibrations ***displacement*** is defined as the distance a point on a machine travels from the top to the bottom of its vibration cycle [11, 16]. This is called displacement "peak-to-peak" (pk-pk) and is most commonly used together with displacement RMS (root mean square that is about 0.707 times pk-pk). Unlike this, ***velocity*** is defined as "the rate of change of displacement". Assuming a constant displacement, the resulting velocity amplitude depends on the vibration frequency. For example, if a peak-to-peak displacement remains at 0.01 mm, the resulting velocity ($v = s_{0-\text{peak}} \cdot 2\pi f$) at 1,800 rpm will be 0.94 or almost 1 mm/s. At 3,600 rpm, the same 0,01 mm displacement will produce 1.88 or almost 2 mm/s. The changes in velocity vary directly with the changes in frequency. But, lowering the speeds, a problem with judging machine health is created.

If we assume that the vibration attentive velocity amplitude is 8 mm/s, at 3,600 rpm, 40 μm will give 7.5 mm/s. At 1,800 rpm it will take 80 μm, at 1,200 rpm it will take 120 μm, and at 1,000 rpm it will take 127 μm.

Now consider low-speed machines (speeds lower than 600 rpm). For an "alert velocity" of 8 mm/s, on a 600 rpm fan or roll, it will take almost 254 μm. A cooling tower fan running 300 rpm will take 508 μm! A 508 μm orbit due to rotor unbalance or shaft/coupling misalignment will result in large amplitude at the gear box's gearmesh frequency. The velocity units work well for gearmesh frequency, and therefore lead the analyst to think there is something wrong with one of the gears. Instead, the gearmesh trouble actually originates from the large displacement orbit that does not seem too bad in velocity terms. Interesting examples were given by [16] describing that, for example, in a paper mill, a large diameter dryer roll running 100 rpm will require an outstanding 1,448 μm before the velocity reaches 8 mm/s. Paper quality can be negatively affected with a displacement of 450 μm, which could lead an inexperienced analyst to think the velocity amplitude is safe enough as it is—only 3 mm/s.

It seems that such low-speed machines show too many problems as indicated by the vibration at gearmesh frequency (number of teeth × rpm). As there are many teeth on large gears, the velocity reading will be read accurately. No correction factor is needed as the gearmesh frequency is usually several thousand cpm (where velocity is an accurate indicator). But the trouble (root causes) would probably originate from the large displacement orbit of the shaft's centerline around the axis of rotation and this can be omitted.

Investigating the best descriptors for low-speed machines within [22], measurements have been performed on drive train of a rotary kiln consisting of drive gearbox, coupling, drive shaft with two plain bearings, and open gear transmission (pinion and girth gear). Measurements were carried out on the two plain bearings with a low-frequency accelerometer. Measurement conditions are given in the Table 3.1.

The measured equipment is shown on Fig. 3.4. Results obtained have shown very low values of all three vibration parameters, the acceleration, velocity, and displacement because the whole system was just before the measurements refurbished and in good operational condition.

But, during operation cases raised when there were uneven coatings in the kiln causing deformations of the girth gear. It has affected the engagement of the pinion and girth gear causing bottoming of the two gears.

Table 3.1 Measuring conditions for exemplary drive train of cement rotary kiln

Measuring items and equipment	Characteristics
Drive gearbox characteristics:	315kW, transmission ration 55.87, n_1 = 1500rpm; n_2 = 27rpm:
Elasting coupling	RWN1120
Drive shaft equipped with two plain bearings	Two-sectional journal plain bearings
Pinion/ Girth gear	$Z_1 = 21$, Do_1 = 693mm $Z_2 = 184$, Do_2 = 6,072mm
Measuring equipment	SPM Leonova, CONDMASTER software
Measuring sensor	Type: low frequency accelerometer

Fig. 3.4 Measured drive train

Still, velocity and acceleration values were decreased, and displacement was the only parameter being significantly increased showing impact of the inconsistent operation of the open drive transmission.

This is given in the diagrams on Fig. 3.5 where marked data are showing the increase of the displacement in contrary to decrease of the vibration and acceleration (data of 13.12.2010). Decease of the operational speed is not the cause of displacement change as the next results with decrease of the speed, all three vibration parameters were decreased (see measurements from 29.11.2010 to 24.01.2011). Appointed date of 13.12.2010 was exactly the case with significantly uneven coatings in the kiln.

Following the above given examples, at low-speed machines, as many industrial analysists noted, it is very important not to relay sole on the estimated parameters from processed data. It is the person in charge who is doing analysis important to "look for things that should not be there" as N. Williams is writing in his work [23] and first of all to know what to look for. Especially in the vibrodiagnostics methods, where failed bearings for example seldom generate high amplitudes in low-speed machines, the analysists should know the expected frequencies and look for any, even minor changes in their amplitude and changes in the time waveform. It coincides with the main rule of monitoring process parameters: follow the trends of certain, pre-known variables.

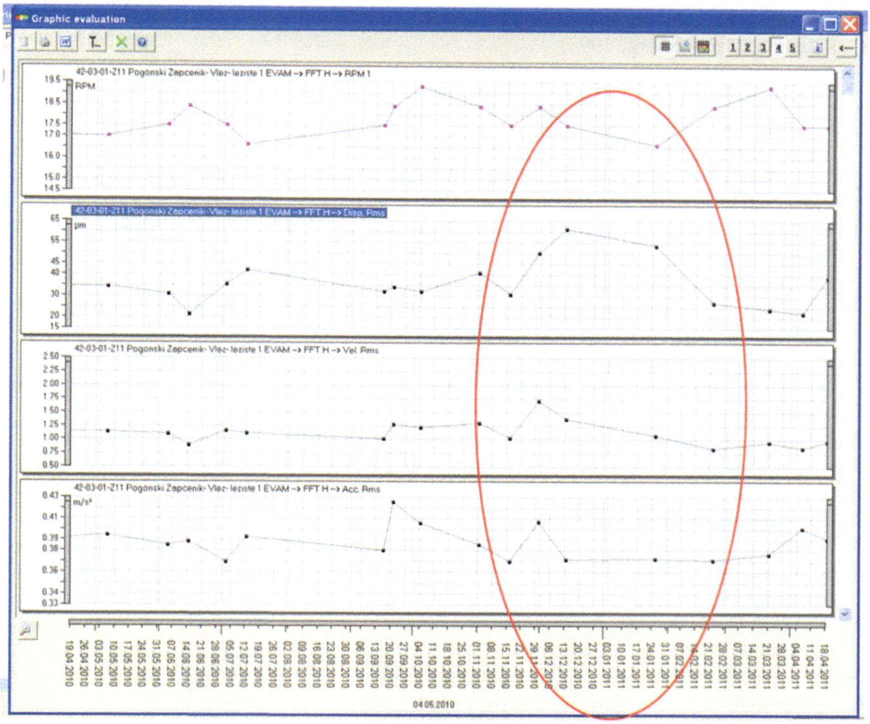

Fig. 3.5 Measurement results (rpm, displacement, velocity, and acceleration) of drive train

In general, low-speed machines are complex for maintenance. The situation is becoming more compound at some machines when a combination of components exists where some are low-speed and some are running with at least medium speeds. Such combination makes their operation and monitoring even more complex, so knowledge on the machines components and operation is more than necessary when designing condition monitoring system.

References

1. A. Jablonski, T. Barszcz, Validation of vibration measurements for heavy duty machinery diagnostics. Mech. Syst. Signal Process. **38**, 248–263 (2013)
2. J.R. Ottewill, M. Orkisz, Condition monitoring of gearboxes using synchronously averaged electric motor signals. Mech. Syst. Signal Process. **38**, 482–498 (2013)
3. W. Bartelmus, Editorial statement. Mech. Syst. Signal Process. **38**, 1–4 (2013)
4. J. Mais, *Low Speed Analysis: Vibration Monitoring of Low-Rotational Speed Applications Using Accelerometers* (SKF Reliability Systems, San Diego, 2003)
5. SKF Reliability Systems, CM1006: condition monitoring ensures against mechanical failure
6. SPM: ILearnVibration, ILearnInteractive.com

7. R.M. Barrett, *Low Frequency Machinery Monitoring Measurement Considerations* (Wilcoxon Research, Gaithersburg, 1993)
8. T. Sundstrom, *An Introduction to the SPM HD Method* (SPM Instruments AB, Strängnäs, 2010)
9. R. Klubnik, *Measuring Displacement Using Accelerometers* (Wilcoxon Research, Germantown, 2008)
10. C.M. Harris, A.G. Piersol, *Shock and Vibration Handbook*, 5th edn. (McGraw-Hill, New York, 2002)
11. A. Ilievski, K. Angjusev, D. Korunski, Z. Petrovski, V. Gavrilovski, Vibracii na masini, pricini, dijagnostika i sanacija (Masinski fakultet, Skopje, 2004)
12. J.P. Den Hartog, *Mechanical Vibrations* (MIT, New York, 1984)
13. ISO10816-1, Mechanical vibration—evaluation of machine vibration by measurements on non-rotating parts-Part 1: general guidelines
14. ISO10816-3, Mechanical vibration—evaluation of machine vibration by measurements on non-rotating parts-Part 3: industrial machines with nominal power above 15 kW and nominal speeds 120 r/min
15. ISO10816-5, Mechanical vibration—evaluation of machine vibration by measurements on non-rotating parts-Part 5: machine sets in hydraulic power generating and pumping plants
16. R.T. Buscarello, *Practical Solutions to Machinery and Maintenance Vibration Problems*, 4th edn. (Update International, Boulder, 2002)
17. A. Barkov, A. Azovtsev, A new generation of condition monitoring and diagnostic systems, a collection of condition diagnostics papers on the Internet cite (1997), http://www.vibrotek.com/ref.htm
18. W. Wier, D. Trivanovic, *Using Acceleration Enveloping on Sleeve Bearings* (SKF Reliability Systems, CM3093)
19. B. Saxer, B. van den Heuvel, Dynamic problems of girth gear drives for ball mills in the cement industry. in *31th IEEE Cement Industry Conference*, Denver, 1989
20. G. White, *Diagnosing a Low-Speed Gearbox Problem. A PREDICT Case History* (DLI Engineering, Washington, 1998)
21. *Low Frequency Vibration Measurements on a Compressor Gear Set* (Wilcoxon Research, Germantown)
22. Z. Stamboliska, Proactive method of low-speed machines condition monitoring. Ph.D. thesis, Wroclaw University of Technology, Wrocław, 2011
23. N. Williams, Slow speed bearing analysis—monitoring in the grey zone. Maintenance Asset Manage. **23**(3), 28–34 (2008)

Proactive Condition Monitoring of Low-Speed Machines

4

Abstract

This chapter represents the principle of proactive approach in condition monitoring (CM) with a focus on low-speed machines. Comparison is made to the standard recommendations for CM, and the weak points of this approach when dealing low-speed machines are highlighted. The theory of a new concept for proactive CM is given.

Keywords

Proactive · Condition monitoring · Low speed · Machines · Preparatory activities · FEM · Tailor-made analyzing tools

4.1 Basics of the Condition Monitoring in Maintenance

Most failures give some warning before they occur. According to ISO13372, failures are defined as equipment condition reaching an unacceptable level, or termination of ability to perform required function [1]. The kind of prior warning given by the equipment is known as "**potential failures**", and one of the best definitions is given by Moubray [2], who defined it as an *identifiable physical condition, which indicates that a functional failure is either about to occur or is in the progress of occurring*. This is the state when the machine can still run, but something is wrong with it. For example, a bearing can reach a temperature of 85 °C in a large gearbox. The machine is still running, but something is incorrect with the bearing and this is actually its alarm limit. If left uncorrected, the machine will reach a state where it cannot perform its operation any more. The other term

Z. Stamboliska et al., *Proactive Condition Monitoring of Low-Speed Machines*,
DOI 10.1007/978-3-319-10494-2_4

that was used "**functional failure**" or "break-down", is the state defined as the "inability of an item to meet a specified performance standard" [2].

Techniques to detect potential failures are known as "on-condition" or **CM maintenance tasks**, because items are inspected and left in service "on the condition" as they still meet specified performance standards. The frequency of these inspections is determined by the "P–F interval" (Fig. 4.1) [2], which is the interval between the emergence of the potential failure and its decay into a functional failure. This time is also known as ***Failure Developing Period*** (***FDP***). This characteristic is vital for determination of inspection frequency. The rule of thumb is to make inspection frequency equal to FDP/2, otherwise the event of failure can occur and its developing period to the break-down will be recognized too late to be able to do something. If a failure is detected early, a break-down and unnecessary downtime can be avoided by scheduling the corrective maintenance for the next shutdown. Having these terms in mind, we can define the **CM as a component** of the Maintenance strategies *providing close monitoring of the vital properties of certain equipment that can indicate the potential failures at the earliest stage of development if not even in the initiating phase.* According to ISO13372:2004 (E), it is about detection and collection of information and data that indicate the state of a machine [1].

Every equipment has certain typical parameters that can be used for monitoring its condition. These parameters can be temperatures of bearings, vibration of the rotary elements, thickness of shells, noises of moving elements, condition of lubrication oils, changes in colors, dust emissions, etc.

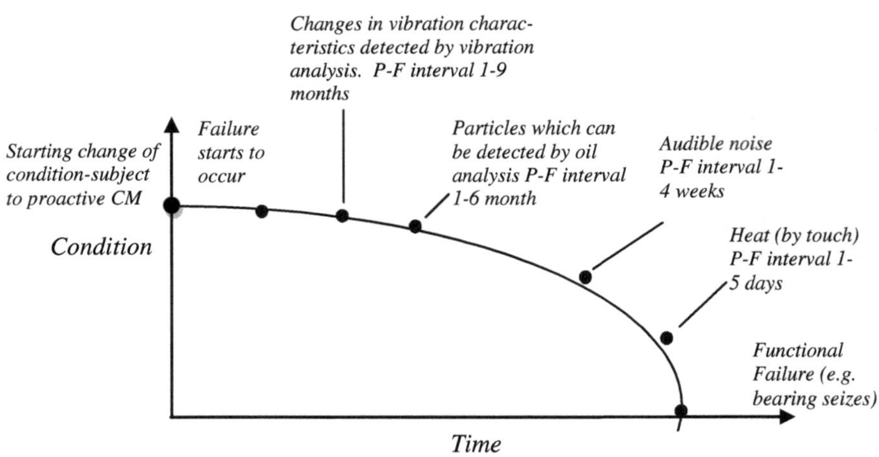

Fig. 4.1 Potential failures and how they can be depicted from condition monitoring techniques [3, 4]

When focusing on proactive CM, the main actions are in point of time located on the Y-axis (Fig. 4.1) or the "zero time", the point when condition starts changing, but is still ahead of a potential failure to start occurring. This is important to notice in order to understand the major difference between proactive CM and typical CM as discussed in detail in the following sections.

4.2 Proactive Approach in Condition Monitoring

In Chap. 2, different maintenance strategies were discussed and among other aspects, it was pointed that the centerline of today's developments is still RCM where proactive activities have the major role as they enable not only to solve problems in advance, but also modification and improvement of the overall condition and thus prolonging equipment lifetime. As defined in EN13372:2004(E), proactive maintenance emphasizes the routine detection and correction of root cause conditions that would otherwise lead to a failure. The **main properties** of proactive maintenance strategy are:

- enabling proactive actions and behavior within maintenance and operation units and
- establishment of CM methods/techniques that would give early recognition of any parameter variation pointing to machinery condition change in a stage where preventive actions are feasible. Detection only of failure effects is not of interest.

According to Industrial/Organizational Psychology, **proactivity** is about being anticipatory and taking charge of situations. It refers to change-oriented and self-initiated behavior that involves **acting in advance of a future situation**, rather than just reacting. It means taking control and making things happen rather than just adjusting to a situation or waiting for something to happen. The use of the word ***proactive*** was for long time limited to the domain of experimental psychology, but from the last decades it is widely used in many other scientific fields. One of the best explanations is maybe the one given by the Austrian neuropsychiatrist, Dr. Viktor Emil Frankl, who used the word ***proactive*** to describe *a person who took responsibility for his or her life, rather than looking for causes in outside circumstances or other people.*

Hence, presenting proactivity in maintenance terms, is about: looking into real root causes of a problem and acting in advance to a future situation by taking responsibility for what is happening to the asset.

The **role of CM** to detect and collect information and data that indicate a change in the state of a machine **when applied for proactive maintenance** has to be designed in a way that would **focus on detection of root causes** of different states of the machine, and not only indicate a change that has already taken place.

In general, **CM** is the process of monitoring a parameter of condition in machinery, such that a significant change is indicative of a developing failure. CM activities follow ISO17359:2003 (E), directed toward identifying and avoiding root cause of failure modes. The use of CM allows maintenance to be scheduled, or other actions to be taken to avoid the consequences of failure, before the failure occurs. Nevertheless, a deviation from a reference value (e.g., temperature or vibration behavior) must occur to identify impeding damages. Once a defect has been identified, the failure process has already commenced and CM systems can only measure the deterioration of the condition that could shorten the normal lifespan. However, CM enables addressing affected conditions before repeated failures occur and such intervention in the early stages of deterioration is usually much more cost-effective than allowing the machinery to fail. In Fig. 4.2, the role of the CM as a component of proactive maintenance is presented. In this manner, for proactive maintenance, the essential care is still assumed as a basic precondition to preventing failures. If all listed "care" activities were performing regularly with success, the number of failures would dramatically reduce. However, failures still

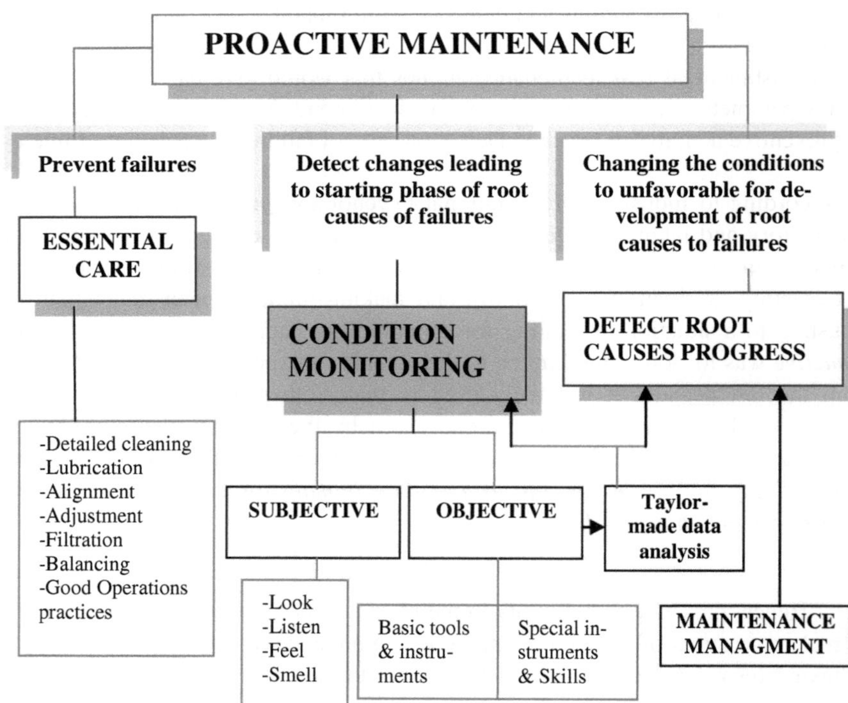

Fig. 4.2 Role of condition monitoring as a component in proactive maintenance

happen and the maintenance aim is to detect and correct them, but as already mentioned, proactive maintenance is the strategy posing additional effort to improve so that a failure could be avoided. In this strategy CM is not used only to detect failures and correct them. Here, the **CM gets new meaning because its role is more to detect any significant changes in the machinery that could be linked with potential failures**.

All findings obtained with any of the subjective or objective techniques are used so as to undertake precorrective actions in the following two domains:

- equipment (changes in the "hardware" such as changes in lubrication systems, oil quality, reinforcement of structures, improved material quality, etc.).
- operations (change of exploitation process and parameters, such as varying the speed of a fan in different speed ranges, redirect gases by regulating flaps, redistribute thermal loads of a kiln, change lubrication regimes, etc.).

The objective findings are analyzed with special, tailor-made data processing tools and in optimized time some extra corrective actions could be undertaken before evidence of failure development. Taking this into account, CM has actually a leading and advanced role in proactive maintenance. It is due to these reasons that the term ***proactive CM*** was introduced in [5, 6].

The scope of proactive CM is a set of CM activities of a machinery to recognize any condition change that could be linked with initiation of root causes of potential failures, and thus serving as a trigger for undertaking early corrective actions to modify equipment or change process parameters in order to move from potentially "dangerous" operating zones, and by this improving and extending the life of the machinery components.

To make this approach realistic, some major differences from today's standardized CM have to be applied in a plant's system. The general recommendations when designing CM program according to ISO13373-1:2002 (E) are to set data acquisition in operating conditions to **approximate the normal operating conditions** of the machine to ensure the consistency and valid comparability of the data.

But what is missing in this standardized approach? It is that in this way, states when faults are actually generating are omitted! Depending on the complexity of the machines, very often exactly in such cases of nontypical operating states, machine components suffering higher loads lead to change in their properties (higher wear rate, deformations, higher stress in the materials, material structure change due to thermal influences, etc.). And these changes are in most cases the actual root causes, as it is a set of conditions that occur at the beginning of a sequence of events that could result in initiation of a machine failure mode.

All of the mentioned change in properties (wear, deformations, stress, material structure change, etc.) can lead to different failures starting from simple bearings defects, up to catastrophic breakdowns of entire machine vital parts.

In low-speed machines generation of faults is more sensitive to the change in operating conditions, as due to low speed, regular changes are at very low rate and can be detected/prevented in most cases with preventive or predictive maintenance actions. Already in Chap. 3 it was elaborated the peculiarities of low-speed machines.

Having the two previous commentaries in mind, and aiming to make practical use of the proactive approach to CM, comprehensive investigation was performed in [5] on low-speed machine failure modes, their **root causes**, the **nature of failure mechanisms, and physical consequences** on the machinery components. The main effort was to find actual answers to the following questions:

1. What are the interrelations between the failure modes and a machine's components condition change?
2. What faults are generating in the machine before failures are to happen and what are the root causes for their generation?
3. What change in the machine condition can recognize a fault pointing to generation? (e.g., how a bearing is behaving, how the wear rate is changing, etc.).
4. Can we measure this change?
5. Can the measuring system be established without large costs and can it be effectively implemented also in the existing controlling system of plants?
6. Can we get final parameters that are easy to be understood by the operator or maintenance staff in this stage of faults generation?

The first two issues can be solved by detailed analysis of the machine design, exploitation conditions, function performance, and operation sequences. To resolve the third issue, see below.

To recognize certain changes when a fault is generating, a special CM program has to be designed that will be able to capture all operating conditions, not omit them when faults are creating, and give data that would enable detection of the root cause subsistence. In this way such CM program would differ from the one designed according to EN13372.

Such a special CM program was developed in [5] and called ***Proactive Low-speed machines CM Method*** (***PLCMM***). The method was developed for low-speed machines as the group of machines with particular sensitivity to operation[1] conditions.

It is actually about monitoring the equipment condition from a new aspect, different from conventional monitoring methods and techniques. It uses the process consequences from the day-to-day operation, treating them as possible root causes for changes in the equipment components' mechanical or electrical condition. It brings together the new trends in proactive and operators-driven maintenance philosophy as organizational aspect and new, nowadays easier available techniques for low-frequency vibrations monitoring as the main CM tool.

In order to point out the differences in the new method from the valid standards, in Fig. 4.3 a CM procedure flowchart is presented according to ISO17359:2003 (E) where new traits differing from PLCMM are given.

The reasons for the changes made are as follows:

A After identifying failure modes, effects, and criticality of a machine, with the new method it is also foreseen to identify possible faults and their generation mechanisms, which are to be treated as root causes. When conducting FMECA, in-depth analysis and investigation should be performed as additional stages on faults generation conditions. Separate diagrams can be made as shown in the example of the cement rotary kiln in Chap. 2.

B Further to this, changes in the machine condition resulting directly from the root causes should be determined and it is most important to recognize if these changes can be measured. This is exactly what the new **method asks to be measured: changes in the machine condition pointing to faults generation and which can be linked to the root causes**.

C After setting and performing measurements, parameters should be defined that will be easy to understand by operators, which in many cases shall be tailor-made descriptors, but would also point directly to some typical condition/state of the machine.

D Use the parameters under C as triggers for immediate actions by the operator or maintenance staff (if applicable) to prevent further faults development.

4.3 Theory and Concept of Proactive Condition Monitoring

The theory is based on the following two items:

- The **natural phenomena of the equipment operation** like the processes of thermal expansion of the equipment components, loads from the production throughput, coatings distribution inside an equipment, eccentricity generated by

[1] For this purpose "operating" conditions assume conditions governed by the process and also by the machine operation in terms of mechanical or electrical properties change.

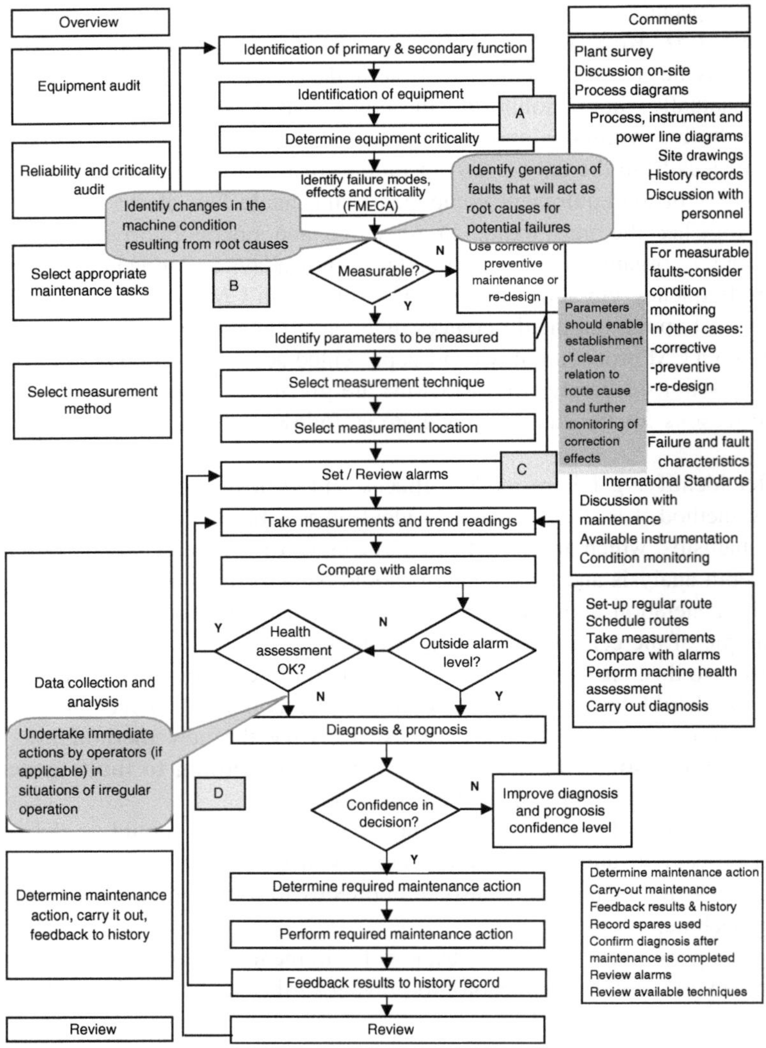

Fig. 4.3 New properties of proactive CM method in relation to typical CM procedure according to ISO17359:2003 (E)

thermal or mechanical cranks, regimes of lubrication at operation of plain journal bearings, etc.; and

- **The measurable or estimated, condition variables changing in time as consequence of the previous**. Here, it is considered parameters described by displacement of some components (as radial run-out or wobbling of rotating elements), temperature (temperature of the bearings and shell temperature distribution), strain in particular sections of structures or other specific parameters, deformation of structures, shock pulses energy, etc.

Proactive CM deals with the later items, the measurable condition variables, because their change points explicitly to the possible condition faults and are good triggers for undertaking certain actions that will correct the root causes and thus eliminate future possible failures. Some of these live parameters change very slowly in time and usually require actions that can be well planned in long time frames. However, some of them may require immediate action because machines operation in the conditions recognized by the limits set for these parameters may bring them down in a very short time.

The concept of the method is shown in Fig. 4.4. It consists of two entities:

1. Preparatory activities,
2. Ongoing cycle of activities.

Preparatory activities are carried out when the method is prepared for application to a specific machine. The main goal of this set of activities is to determine the failure modes, faults that can be generated and in what way, determine representative measuring parameters for the faults, and link the change in these parameters with possible corrective actions for faults prevention.

When establishing CM system for proactive maintenance purposes, the first thing is to decide if this type of CM is really what the plant needs, if the company is ready to accept it and use, and last but not the least, the company is really determined to implement it. These questions are to be decided at the upper managerial level and once agreed, all the tools of management change should be put in place until the approach becomes really "life".

There are two general ways to implement this system:

- Implement it only for major and critical equipment; and/or
- Implement it as regular proactive philosophy in equipment maintenance.

The first case is easier to be established and it needs a systematic way of carrying out necessary steps whose effects are easily visible.

The second case is much more difficult to be established. It requires change in the company's culture in how things are handled. A company cannot have different ways of handling production process and maintenance issues. It requires a longer time than the first way, great effort by the company's top management, strategies, and effective change management system. However, once established, the results would be visible in all aspects as safety, environment protection, operations processes control, maintenance, optimization, etc. It would mean a change in all employees' mentality and acting proactive in all aspects of their life too. In this book, the focus is put on the first, systematic approach as it can be described and implemented by following certain technical steps.

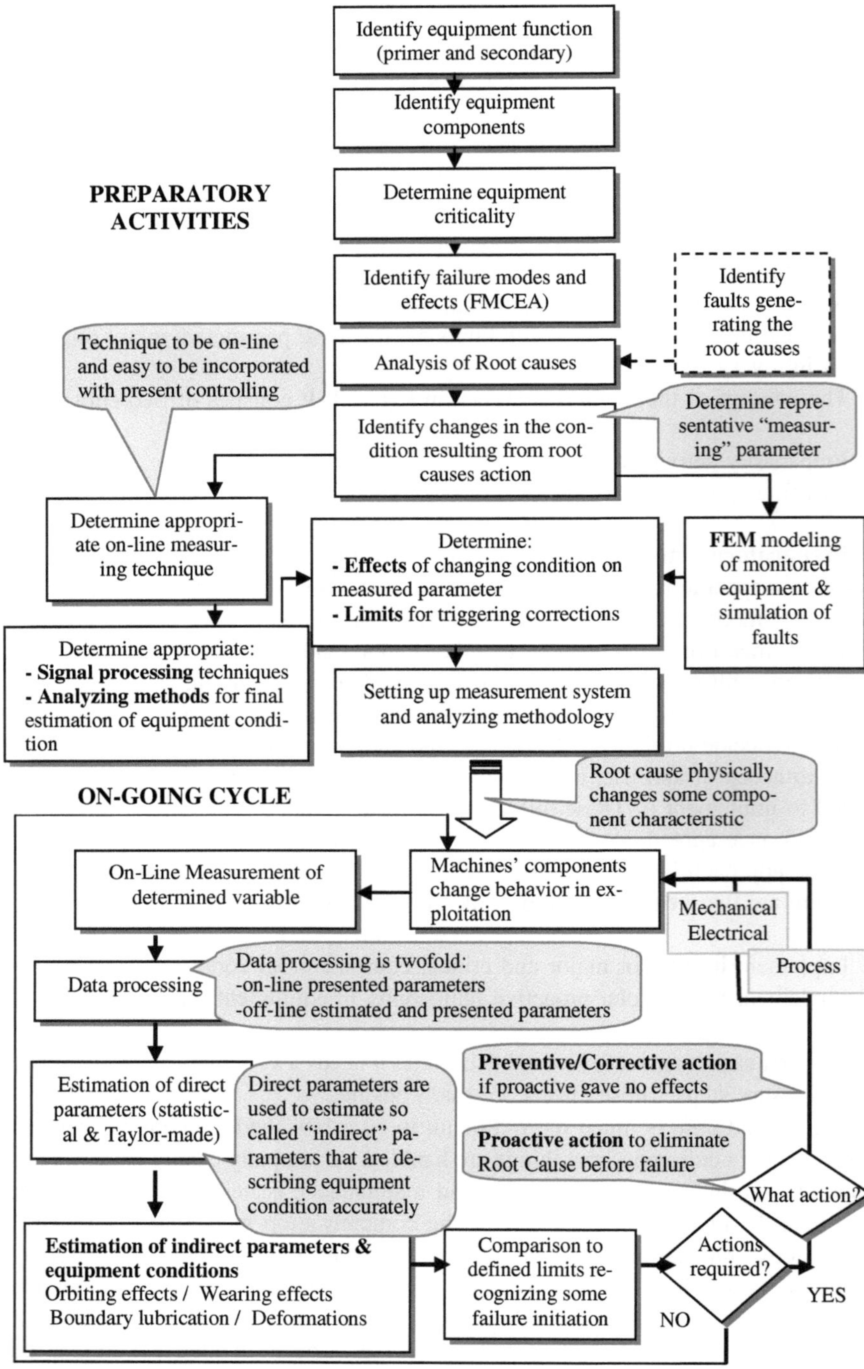

Fig. 4.4 Concept of the new method for proactive monitoring

4.3.1 Preparatory Phase

Preparatory activities should be carried out in a systematic way following all the given steps as they put the foundations for application of the whole method.

Step 1 Identify primary and secondary functions of equipment as both are critical for determination of criticality, as well as for the further measuring steps and reacting actions. A good practice is to list them with a short description and the associated risks.

Step 2 All major components of the equipment should be identified as separate components may require separate CM methods. For example, in a ball mill, the body and the drive train are separate components of the machine and they require very different monitoring methods and techniques.

Step 3 Determine equipment criticality as a whole unit, but also of the separate major components, depending on their function. Widely accepted criteria for criticality can be used.

Step 4 Identify failure modes and effects using FMECA and giving high focus to high-risk modes, in particular to the present possibilities to detect them in their initiation phase.

Step 5 Analysis of root causes and the factors affecting them, the faults that initiate them, and their early symptoms.

Step 6 Identify changes in the condition of the equipment/component affected by the root causes. This is an important step as it determines the representative measuring parameter(s) that are to be used in the proactive CM. In this case, a representative parameter can be a temperature of a bearing, shell, or other component because it is influenced by the friction process, contacts, lubrication regimes, etc. Pressure in the fluids, pressure of gas flow or state in some chambers can also indicate if there are any irregular changes going on. Change in the position on macro level or displacement of some components having limited freedom to deflect under increased load or under effect of thermal load or due to increased deformations can in many cases point to some starting generation of faults in the assemblies. In low-speed machines as discussed in the chapter for their maintenance, a part of parameters as temperature, pressure, current, etc., the displacement is the best descriptor of their condition (see Chap. 3) in most cases. Therefore, attention should be paid to what displacement, where to be measured, and how to be measured (relatively to some machine component or to the fundamentals). When selecting measurable variable, it should be well thought out how to select appropriate and meaningful displacement. Here we can consider radial displacement of trunnions at ball mills, large shafts and plain bearings, shafts of drive trains, especially where meshing of pinion and girth gears can be monitored. Also, it is important to note and take into consideration the relativity of the measurement, e.g., measured displacement changing relatively to which other components of the machine. The further actions go in parallel:

Step 7a Determine appropriate online measuring technique, but together with signal processing and analyzing methods for final estimation of the equipment condition.

The measuring parameter also determines the measuring system to be installed. Measuring system must take into account the exploitation conditions because in heavy-duty industry often harsh environment conditions are present, such as high temperature, diffuse dust, exposure to ambient conditions, oil leakages, vibrations being transferred from different surrounding machines, etc.

Based on the research [7–9] and also the experience in industry, eddy current probes have proved to be the most suitable probes for low-speed machines, lower than 10 cpm, even less than 1 cpm. However, there can be more different sensors applied such as piezoceramic accelerometers, proximity (eddy current), electrodynamic velocity sensors, and Piezo-Velocity transducers (PVT) or shock pulse transducers, as already described in Chap. 3. The measuring probe should be best suitable for the measured property and conditions of measurements.

In this phase also methods and tools for signal processing should be selected or designed, and special focus should be on analyzing tools. Measured data should be available in the control rooms (CR) in the form that would correspond to the staff skills for understanding their meaning. The next step is to decide on what estimated parameters would be used for further estimation of the monitored equipment condition. **In order to come to such parameters, it is in most cases necessary to develop special, tailor-made analyzing tools**. These tools can use:

- Direct visualization of the change in measured parameter in time with defined limits for normal operation. Visualization should provide clear information for the actual readings, enable follow-up of trends of parameters change, but also enable zooming for some particular segment of interest;
- Values of statistical parameters S_{MIN}, S_{MAX}, $S_{peak\text{-}to\text{-}peak}$, Standard deviations, and filtering for extraction of amplitudes or events of interest;
- If dealing with vibration-related parameters, then FFT spectrum with predetermined harmonics of interest and limits set for the equipment components are to be included. Harmonics should be related to the measured machine components, but also to engaged components, like for example gears, tyres, etc.

Analyzing tools are much dependent on the equipment to be monitored because they have to be connected to the geometry of the machine, principle of operation and last, but not the least, the nature of the operational processes in order to create the link between the measured parameter and the physical changes in the equipment.

Step 7b In parallel to the latter, it is required or at least it is very beneficial to **design FEM model** of the monitored equipment. The aim is to provide **simulation of the exploitation conditions** and determine the effects, considering also the worst case scenarios that can happen in real exploitation due to extreme, but realistic loads (mechanical or thermal) and set safe limits for the monitoring method and equipment. For such heavy-industry cases, real exploitation cannot be used to determine these extremes because such "experiment" would be too costly. Therefore, creating FEM model and simulating the operational conditions, stresses in the separate components can be estimated that could be used for estimation of the lifetime of the components and setting limits that have to be applied in the real equipment in order to avoid failures. Even more, limits can be connected to phenomena such as deformations, loads, temperatures, etc., but they can be translated into values of the measured parameter (e.g., displacement) thus making the function of the method easy to understand for end-users.

One of the important actions at this stage is the determination of measured parameters and possible changes:

- Ranges of the monitoring parameters should be defined for the condition of the equipment accepted as normal or satisfactory;
- Changes in the parameters should be defined in a way to be easy to bring a conclusion that something is about to occur within the equipment; and finally;
- All limiting values of parameters should be linked to possible initiation of faults in a way easy to be understood by the users.

Step 8 **The next step is to determine what actual** conditions (states) can be recognized through the estimated indicators based on the measured variables. Depending on the equipment, the following examples of nature of faults can be recognized:

- Distribution of the displacement within the MIN–MAX values for one revolution with very low speed represented in an "orbit" plot shows the "**out-of-roundness**" of the rotating component, but also all deviations taking place at some position (contacts, deformations, nonuniform wear, etc.).

- Deviations in the equipment reference position, projected by the magnitude of the harmonic corresponding to the equipment (or component) revolution in the frequency spectrum of the displacement. FFT can point to development of **radial run-out** and development of some **eccentricity or deformations**. If these values remain for a longer period, the temporary change may lead to **permanent deformations**.
- Analyzing the magnitude in the same FFT, information for journals or trunnions "**orbiting**" can be obtained. Based on this, wear trend of the Babbitt bearing can be estimated and actions planned for reducing the wear rate.
- Changes in the relative position of some components regarding others, for example change in position of journals regarding bearings bushing can point to **wear of bearings in long terms**, but also in short terms it can point to **change in the lubrication regime** due to clearance changes in the bearing assembly.

Overall, all listed parameters and their progress over time in the system of online measurement gives the possibility to compare their values to some given limits. Every tendency to approach the limits can serve as a trigger to react early and reduce significantly the possibility of creating the state of the equipment where some of the previously mentioned final failures can occur. The reactions, as specified, could be process or mechanical.

Step 9 The final step of the preparatory phase is actually to set up the measuring system and analyzing methodology at the users (install and commission the equipment and software, train the operators). Based on the parameters' nature and exploitation terms of the equipment, proper measuring techniques and tools should be selected. They should be incorporated into adequate monitoring system and set proper inspection intervals as well as methodologies for evaluation of the received results.

During the preparatory phase, a set of procedures should be defined that will be used by the operators and maintenance staff as reaction actions to the alarms generated by the monitoring system. Procedures or actions should be clearly defined to be undertaken in proper time to prevent functional failures and even more, to improve the condition of the equipment because of possible potential failures

4.3.2 Operational Phase

The set of activities in the **ongoing cycle** represents the functioning of the method on-site. It requires previous completion of the preparatory activities, install the measuring system, and analyzing tools and training of the operators on the method principle, required process corrections, train maintenance staff, etc. The system should be continuously functioning and measured data and estimated parameters presented to the end-users in real time as the main property of the approach was exactly to recognize all deviations from regular state. This means to set precisely the sampling rate of the measuring system and have communication and software applications available to operators and/or maintenance staff. The estimated direct parameters can be presented or they can be used only to estimate the indirect, but physically meaningful parameters. The latter can be related to orbiting effects of any rotating component, deformations, deflections of structural elements, conditions for boundary lubrication, vibrations of interest, etc. Here, the role of the previously defined limits with the use of FEM comes into play. Actually, although the condition of an equipment is satisfactory, one set of the limits can be placed in a way to recognize the start of a fault initiation. This will alarm the plant to undertake proactive actions to eliminate the root cause before a failure mode is established. For example, such actions could be changing the regime of operation of a machine (decrease the speed, throughput, fuel, etc.). In most cases the first reactions are process reactions, which are also called "operators driven" maintenance activities. It means that operators should undertake the first activities to do some process changes and change the negative trend in the change of equipment reliability. Here, software can be a helpful tool to help operators in which other process parameters to check, and make quick decisions (see the details in the example given in Chap. 7).

Mechanical and electrical activities should come at the "end", when the operation cannot improve the process any more (thermal or mechanical loads, flows, capacities, etc.). Usually, these are more extensive corrections starting from the simplest like forced external cooling, realignments, etc., up to complete disassemblies, corrections of mechanical components, redesign part of the equipment, apply new materials, oils, etc.

In Chap. 7 this method will be further worked out by applying it for monitoring of an exemplary low-speed machine: cement rotary kiln. Details of the method will be adopted for this application by defining reliable measuring parameter, designing measuring system, designing signal processing, and analyzing tools and setting alarms for triggering corrective actions. FEM will be also applied for definition of limit values of the monitored parameter as their determination through practical cases is not recommendable as it would mean huge defects in the plant.

References

1. ISO13372 Condition monitoring and diagnostics of machines-Vocabulary
2. J. Moubray, *Reliability Centered Maintenance*, 2nd edn. (Industrial press Inc., New York, 1997)
3. Preventive Maintenance/Essential Care and Condition Monitoring, IV edn. (IDCON, Inc., 2003)
4. G. Pantazopoulos, *Operationa and Strategic Planning for Continuous Improvement of RCC* (Technical Department, Roanoke, 2002)

5. Z. Stamboliska, Proactive method of low-speed machines condition monitoring, Ph.D thesis, Wroclaw University of Technology, Poland, 2011
6. E. Risinski, Z. Stamboliska, P. Moczko, *Proactive Control System of Condition of Low Speed Cement Machinery, Automation in Construction*, vol. 31. (Elsevier, 2013), pp. 313−324
7. M.B. Richard, *Low Frequency Machinery Monitoring Measurement Considerations* (Wilcoxon Research Inc., Gaithersburg, USA, 1993)
8. M. Jason, *Low Speed Analysis: Vibration Monitoring of Low-Rotatonal Speed Applications Using Accelerometers* (SKF Reliability Systems, USA, 2003)
9. M. Jason, S. Brady, *Introduction Guide to Vibration Monitoring: Measurements, Analysis and Terminology*. (SKF Reliability Systems. USA, 2002)

Condition Monitoring Techniques for Low-Speed Machines

5

Abstract

Review of condition monitoring techniques for low-speed machines is given describing their possible usage. Good practices for monitoring and technical assessment of new and already in use equipment for most frequently used low-speed machines in heavy-duty industry are given. How to obtain information for proactive actions from the listed techniques is explained. Special focus is paid on vibrodiagnostics as condition monitoring tool.

Keywords

Condition monitoring · Low-speed · Techniques · Methods · Monitoring practices · Technical assessment · Vibrodiagnostics · FFT · Ultrasonic technique · Geometrical measurements · Strain measurements · Lubrication · Rotary kiln · Ball mill · Excavator · Crane · Crusher · Fan

5.1 Review of Condition Monitoring Techniques

The tools of the Condition Monitoring are two categories of inspections:

- Subjective
- Objective

These two categories are also valid for low, medium, and high speed machinery. However, the importance and the information obtained by them can vary depending on the machinery speed and principle of operation. Therefore, in this chapter the techniques will be described in a manner of what we can get as information particularly for low-speed machines.

Z. Stamboliska et al., *Proactive Condition Monitoring of Low-Speed Machines*,
DOI 10.1007/978-3-319-10494-2_5

Subjective inspections provide information that we can obtain from the four basic senses: **Look, Listen, Feel, and Smell**. Some of the more experienced generations of the maintenance personnel, and specially supervisors should widely use these tools every time passing by the equipment. These senses play very important role in the low-speed machinery as the low moving components allow good visual inspection and overview, as well as sound distinction of the life parts and moving components. The low-speed movements usually produce lower noise that provide better conditions for noticing any irregularities such as dry friction, strokes, knocking, etc. Many useful data can be obtained by the general visual condition of the equipment such as leakages, cracks, broken or missing items, lubrication condition, noises coming from damaged internal parts, higher friction, dry conditions of moving items, higher temperatures of bearings or housings, higher vibrations or strokes, wasted lubrication, gasses leakages, burnt items, etc.

This info is very useful as the first set of information or even as alarming or triggering data for corrections. However, they all cannot be a base for systematic way of condition monitoring. Especially, the strategy of proactive maintenance requires more precise data in order to monitor trends and determine exactly level of certain parameters thus obtaining enough data for improvement actions. From these reasons over the years many kind of equipment condition inspections have been developed for determination of various conditions of equipment.

Objective inspections require special instruments and skilled or trained personnel as well as documented system to gather all the data in historical way and enable detailed engineering evaluation and sometimes even statistical analysis of what was recorded as state. There are many sources on these inspections, but the most important and most frequently applied inspections, which can be classified according to the symptoms or potential failure effects they are focused to, are summarized as follows [1–6]:

- *Dynamic effects*: monitoring detects potential failures, which cause abnormal amounts of energy to be emitted in the form of waves such as vibration, pulses, and acoustic effects of rotational and reciprocating equipment as well as stationary structures or equipment being under influence of external vibrations. Applicable techniques [7–24] are broad band vibration analysis, octave band analysis, real time analysis, time waveform analysis, frequency analysis, Cepstrum, Amplitude demodulation, peak value analysis, Spike energy, Shock pulse monitoring, Ultrasonic analysis, acoustic emission (AE) etc. Application of this kind of inspections for low-speed machinery requires special measuring instruments and estimation rules as dynamic effect differs significantly at medium and high-speed machinery.
- *Particles effects*: monitoring detects potential failures, which cause discrete particles of different sizes and shapes to be released into environment in which the item or component is operating. Applicable techniques [25, 26], are analytical ferrography, mesh obscuration particle counter, all metal debris sensor,

magnetic chip detection, patch test, sediments, etc. These techniques are also used for analysis of different types of oils (lubrication, hydraulics, insulation). They can indicate problems such as machine degradation (e.g., wear), oil contamination, improper oil consistency (e.g., incorrect or improper amount of additives), and oil deterioration. The inspection is equally valuable for low, medium, and high-speed machinery.

- *Chemical effects*: monitoring detects potential failures, which cause traceable quantities of chemical elements to be released into environment. These techniques are mainly applied in condition monitoring for monitoring of lubrication oils and indication of potential failure in the system and not in the fluid itself. It consists of analysis against fluids deterioration and contamination on time determination whether there is oxidation, contact with wear metals, contact with moisture, or any other contamination. It helps recognition of wearing in the gears, bearings, turbines, failure of seals, covers and breathers, filtering appropriateness, etc. Applicable techniques are Atomic emission spectroscopy, AE-rotating disk electrode, Atomic absorption, X-Ray Fluorescence Spectroscopy, Dielectric strength, Infrared Spectroscopy, Thin-layer activation, Electrochemical corrosion monitoring, Moisture monitor, Clear and bright test, Crackle test, etc. This inspection is especially important for monitoring large dimensions low-speed sleeve bearings where the trends of damage progression usually goes very low at the initial stages, but then again at some moment of time it may rise in very short time to final damage and inability of the bearing. The large size of bearings is what makes it very costly and time consuming task for its replacement.
- *Physical effects*: monitoring detects potential failures, which cause changes in the physical appearance, state, or changes in the structure of the equipment which can be detected directly and the monitoring techniques detect potential failures in form of deformations, cracks, fractures, visible effects of wearing and dimensional changes, corrosion, etc. It is applied for examination of solid bodies (structure) of various equipment, welding, pipelines, etc. It enables determination of the quality of the applied materials and their change with exploitation, internal, and subsurface and surface defects (voids, cracks, grains, etc.), changes in thickness, corrosion (active or passive state of corrosion and corrosion extent), etc. Applicable techniques are Liquid dye penetrants, Electrostatic fluorescent penetrant, Magnetic particle inspection, Strippable magnetic film, Ultrasonics techniques, Eddy current testing, X-Ray Radiography, Rigid Borescopes, Deep probe Endoscope, Oil Appearance, Strain Gauges, Viscosity monitoring, etc.

Herewith, a very important for low-speed machines are the **strain monitoring** (loads measurements) techniques for detection of structural health of machines that are of crucial importance due to very low attention paid and catastrophic failures related to it. These measurements can detect potential failures very early and thus give great possibilities for proactive activities for improving the condition of the machine.

Another very important, simple, but valuable in the heavy-duty industry are **Geometrical** (also known as geodetical) measurements applied for examination of the equipment and whole machinery condition at installation phases and changing trends during the years of operation to determine whether the machinery has wearing problem of main components, alignment problems, permanent deformations, or foundation settling problems, etc. This is especially important inspection because with the low-speed of the machines, usually users or even maintenance teams are misled that the condition of the equipment is stable and no physical effect takes place over the time until it is usually too late to undertake proper corrections. This way worn parts can be left un-maintained such as anti-wear shells, liners, supporting rollers (carrying surface), etc. Temporary deformations may go undetected and being transformed into permanent deformations that would than cause alignment problems, bearings heating problems etc.

Lately, there are new measurements known as **Photogrammetry** being widely used for the control of physical effects and so thanks to the modern cameras and laser system. It will be explained later in the chapter.

- *Temperature effects*: monitoring techniques look for potential failures, which cause a temperature rise of the equipment itself. It can detect problems such as excessive mechanical friction (e.g., faulty bearings, inadequate lubrication), degraded heat transfer (e.g., fouling in a heat exchanger) and poor electrical connections (e.g., loose, corroded or oxidized connections). These are based on the principle of emissions of infrared radiation. Applicable techniques are Infrared Scanners, Focal Plan Arrays, Fibre Loop Thermometry, Temperature indicating Paint etc.
- *Electrical effects*: monitoring techniques look for change in resistance, conductivity, dielectric strength, and potential. Applicable techniques are Linear polarization Resistance, Electrical Resistance, Potential Monitoring, Power Factor testing, Breaker Timing testing, Motor Circuit Analysis, Motor current Signature Analysis, Power Signature Analysis, Partial discharge, High-Potential Testing, Magnetic Flux Analysis, Battery Impedance testing, etc. These techniques can help in detection of electrical insulation deterioration, broken motor rotor bars, and a shorted motor stator lamination.
- *Process parameters monitoring or Performance monitoring*: monitoring techniques that can predict problem of various nature by monitoring changes in variables such as pressure, temperature, flow rate, electrical power consumption, and equipment capacity. These parameters are very sensitive to the nature of the technological process and it is of great importance to know the process in details in order to be able to read the process parameters from equipment condition aspect. They can be very helpful for recognition of increased loads, change in components macro-geometry in cases of fluids flow trough changes of pressure over some sections, changes in geometry in machinery exposed to high temperatures due to

thermal loads, etc. It is also the basis of the operators-driven maintenance philosophy when dealing monitoring of process performance trough equipment condition change.

5.2 Condition Monitoring Practices for Low-Speed Machines

Condition monitoring tasks are maintenance techniques that are used to detect the onset of equipment, so if we start from the main target of the condition monitoring: to predict life expectancy of a component than it is evident that monitoring should be focused on the vital characteristics determining the life of a component. It makes no sense to monitor certain characteristics if they do not influence the life in a crucial way directly or help indirectly to determine other related information. In today's world it is far too expensive to spend time and money if the information obtained cannot bring some benefits. When selecting condition monitoring technique, the machinery and its operation should be well known. In order to determine effective/efficient and technically/economically feasible and justified techniques, several factors must be considered:

- Possible failures of the equipment (FMCEA).
- How (if) an onset of failure can be recognized.
- What are measurable parameters for the failure onset.
- P–F intervals of the failures (time between the onset of a failure becomes detectable and the point at which the condition of the equipment deteriorates to a functional failure).
- Required measurement precision and sensitivity.
- Skills of the personnel performing CM activities.
- Resources versus risks.
- Environment, location, and portability.

The possible failures and measuring parameters that can point to failure onset are the basis for the selection of the type of technique (e.g., dynamic effects, particle effects…) in typical (conventional) CM, but in proactive CM one more step is required and the technique and parameter to be monitored should be those contributing to first initiation of potential failures, and not the failure itself.

At typical CM, the P–F interval determines the interval at which a CM task should be performed. The frequency should be sufficient to point the failure onset and give time for reaction. Too high frequency would results in unnecessary costs, while too low frequency can bring inability to react and correct the condition as the onset was detected too late in the P–F interval, or in worse case, the P–F interval could have been even not detected. The consistency of the P–F interval should be considered. If this interval is changing, the shortest interval should be considered.

However, if this inconsistency varies too wide, maybe a CM task would be very difficult to be scheduled or if effective and economically justified, continual monitoring should be employed.

Measurement precision and sensitivity determines the reaction time available to reduce or eliminate the consequences of the functional failure. Some techniques can detect failures onset very early, while others very late. For example, an ultrasonic testing can detect initiation of a crack at undersurface layer thus giving enough time for reaction. Visual inspection would detect the crack, but when already at the surface and in most cases it can be too late for prevention of the failure. Of course, economical side and risks should be also considered when choosing this property of the CM technique.

At proactive condition monitoring, in most cases online measurement systems are required exactly because findings are related not only to the event of failure onset detected, but also to the effects of contributing factors for first initiation that may be an event of short durability that may be missed with periodical measurements (e.g., overload may happen only short time few times per month and be not recorded with periodical monitoring).

Skills of the people performing CM tasks are directly governed by the technique selected. For some techniques a high skilled personnel is required with high specializations. Therefore in some cases, it is better to contract the CM activity with external companies.

The risks related to safety, health, environment, economical losses and other potential losses have to be considered. Effects and the probability of failure occurrence have to be connected and estimate if condition monitoring activity is justified. In some cases, if the impact of the functional failure is less than the CM task itself, than it is not economically feasible to do any CM task.

Environment under which the CM tasks should be performed, the location of the equipment to be monitored and the portability should be analyzed when choosing proper task. In some cases of harsh and dangerous environment, engaging people directly is not justified.

When deciding on the monitoring techniques to be used and their performance, decision should start from the fact if the equipment is:

- Old in operational terms, or stayed idle after supply.
- Newly supplied and put into use.

Next should be considered working conditions, overall maintenance support, technical documentation, and availability of spares. But, the biggest difference in how we establish the process is exactly if the equipment is already long time in use or relatively newly supplied. The basic difference is following:

- *Old equipment*: It is of **crucial importance to assess the technical condition** of the equipment at certain periods of time. Therefore, it is not enough to make baseline measurements only, but it is primary to make detailed checks of the all vital parts of the equipment and based on the findings decide on the next condition monitoring techniques and schedules. It is **beneficial to make also assessment of the remaining life-time** of the equipment with aid of FEM systems, but sometimes it may be costly technique that should be used only for the core equipment (such as mills, kilns, excavators-vital parts, crushers etc.).

- *New equipment*: It is considered equipment to be in good condition (Reliability of the suppliers should be considered). Therefore, it is necessary just to make the baseline measurements (if nonexisting) to picture the starting condition in real-exploitation conditions in order to have clear insights for recognizing faults start developing. Manuals and recommendations from the manufacturers should be followed up and as much as possible implement from the very beginning of the equipment use.

5.2.1 Recommended Condition Monitoring Practices for Running Equipment

Considering the heavy-duty industry and use of low-speed machines, set of measurements, and controls can be recommended for old equipment when doing assessment of the technical condition, while some can be used also for regular condition monitoring.

First thing to do is always check and provide proper functioning of all Safety-related equipment and conditions. This will not be dealt here, but it is always the first thing to be done.

Based on the experience and many literature sources [2, 3, 9, 26–29] the best practices are listed in Table 5.1 explaining also the focus of each technique.

When dealing old equipment, findings from the assessment of the technical condition and general recommendations for inspection of particular equipment should be paired and decide on most effective condition monitoring set of activities. More critical items should receive higher attention than items estimated as being in good condition. Of course, it should all be also related to the criticality of all items for the overall machines and plant reliability.

5.2.2 Recommended Condition Monitoring Practices for New Equipment

When installing new equipment, two main groups of checks should be performed:

- Check of machinery condition against agreed delivery terms (technical specification); and
- Check the machinery against valid standards or best available practices for the corresponding equipment.

Table 5.1 Assessment of technical condition (recommended for equipment in use)

Equipment	Condition assessment	Items to be checked
BALL MILLS	Foundations check (condition of concrete, supporting structure, levels disorder)	▪ Foundations sinking (deterioration); causes ▪ Concrete quality ▪ Oil or other fluids acting over foundation ▪ Health of metal part of supporting structure including anchoring
	Condition of mill body (shell, trunions)	▪ Existence of cracks, voids, laminarities, reduction of thickness or other internal defects in the body of the mill, particularly at the transition areas in trunions and around mill manholes by NDT (Dye penetrant, magnetic particles testing, Ultrasonic tests)
	Condition of trunion bearings	▪ Scratches or wearing at bearings, functionality of lubrication and other hydraulic systems (at starting phases, at operation, filters, etc)
	Condition of drive-train	▪ Open gear transmission: cracks at surface (Dye Penetrant), near under surface (Magnetic Particles effects), deeper in teeth or gear rim (Ultrasonic test); pitting and other defects. ▪ Toothing at girth gears and pinions: backlash and top clearance, radial and axial wobbling, clearances at joints, tightening of connecting bolts, assembly of the gears to the shafts or mill's body ▪ Condition of lubricant and

(continued)

Table 5.1 (continued)

		functionality of lubrication system ▪ Bearings at drive shaft: wearing, sinking, lubrication ▪ Gearbox (regular checks as for any gearbox: gears, bearings, lubrication)
	Overall operational condition by vibrodiagnostics	▪ Overall condition by vibrodiagnostic applied for gearbox, driveshaft bearings, particularly bearings at the pinion. ▪ Look for GMF (gear mesh frequency) & Natural frequency of large gears, their amplitude, sidebands and their spacing and amplitude. Also harmonics and sidebands at running speed of gears. Phase differences of 90 or 180° at bolts, base plates or foot of the components. All these items may help detection of wearing of teeth, misalignment, broken teeth, problems with backlash etc. For more details see Table 5.4. (gears related symptoms) ▪ SPM or similar technique appropriate for roller bearings of low speeds or displacements at plain bearings
ROTARY KILNS	Foundations check (condition of concrete, supporting structure, levels disorder)	▪ Foundations sinking (deterioration); causes ▪ Concrete quality ▪ Oil or other fluids acting over foundation ▪ Cracks in metal construction, anchoring bolts corrosion of drive station foundations, base plates of

(continued)

Table 5.1 (continued)

		supporting stations and rollers
	Condition of kiln shell & tyres	▪ Cracks, voids, laminarities, reduction of thickness or other internal defects in the kiln shell and tyres by NDT. Particularly check the areas near welds at the shell and around manholes ▪ Visual control of tyres against wearing marks, scratches, pitting, spalling, other for surface damages, hot spot or even cracks ▪ Dimensional control of tyres against cone shape of tyres ▪ Visual control against any looseness or cracks in fixation systems of tyres ▪ Review history records of tyres migration (preferably considering also shell temperatures) ▪ Review history records of shell temperature and brick linings
	Condition of kiln axis and deformations	▪ Geometric measurements in cold or hot condition for axis condition (vertical and horizontal deflections) (to be related to foundations conditions also) ▪ Shell deformation measurements ▪ Ovality measurements at tyres areas
	Condition of supporting rollers & thrust rollers	▪ Rollers shafts: internal defects or cracks in transition areas by NDT ▪ Rollers body (carrying surface): *Visual* control of

(continued)

Table 5.1 (continued)

		overall integrity, control against wearing marks, scratches, pitting, spalling for surface damages or even cracks; *Dimensional* control against cone shape of roller and comparison of dimensions of rollers at same supporting station ▪ Control of hydraulic system against functionality and control of axial travelling
	Condition of supporting rollers bearings	▪ Scratches or wearing at bearings, functionality of lubrication and cooling systems by visual control, condition of lubricant ▪ Review history trends of temperatures, deflection measurements (if any) ▪ Scratches at journals (visual, touch)
	Condition of drive-train	▪ Open gear transmission: cracks at surface (Dye Penetrant), near under surface (Magnetic Particles effects), deeper in teeth or gear rim (US); pitting and other defects ▪ Toothing at girth gears and pinions: backlash and top clearance, radial and axial wobbling, clearances at joints, tightening of connecting bolts, assembly of the gears to the shafts or mill's body ▪ Condition of lubricant and functionality of lubrication system ▪ Bearings at drive shaft: wearing, sinking, lubrica

(continued)

Table 5.1 (continued)

		tion ▪ Gearbox (regular checks as for any gearbox: gears, bearings, lubrication) ▪ Effect (if any) of kiln crank on gears meshing at girth gear-pinion toothing ▪ Visual control against any looseness or cracks in fixation systems of girth gear, especially at the connections to shell
	Overall operational condition by vibrodiagnostics of drive	▪ Overall condition by vibrodiagnostic applied for gearbox, driveshaft bearings, particularly bearings at the pinion ▪ Look for GMF (gear mesh frequency) & Natural frequency of large gears, their amplitude, sidebands and their spacing and amplitude. Also harmonics and sidebands at running speed of gears. Phase differences of 90 or 180° at bolts, base plates or foot of the components. All these items may help detection of wearing of teeth, misalignment, broken teeth, problems with backlash etc. For more details see Table 5.4. (gears related symptoms) ▪ SPM or similar technique appropriate for roller bearings of low speeds or displacements at plain bearings
DRYERS	Foundations check (condition of concrete, supporting structure, levels disorder)	

(continued)

Table 5.1 (continued)

	At rotary dryers:	
	○ Condition of shell & tyres	The same as for rotary kilns
	○ Condition of axis and deformations	
	○ Condition of supporting rollers & bearings	
	Condition of drive-train	
	Overall operational condition by vibrodiagnostics of drive	
EXCAVATORS	Inspection of overall integrity	▪ Visual and functional controls ▪ Excavating loads in operation by measuring deflection of supporting structure in 3 directions and use of FEM for calculation of real loads in exploitation and control highly-stressed areas
	Inspection of load carrying structure	▪ NDT testing ▪ Geometrical measurements of large deformations ▪ Load of excavation by use of calibrated strain gauges ▪ On-line monitoring of highly-stressed structure sections by strain gauges against allowed limits ▪ Vibration monitoring of structure against permissible limits

(continued)

Table 5.1 (continued)

	Condition of drives	▪ Overall condition by vibrodiagnostic applied for gearbox higher speed stages ▪ Visual & NDT ▪ Noise measurements ▪ Thermography ▪ Geometrical measurements of gearbox housing deformations causing problems with gears allignment
	Overall operational condition by vibrodiagnostics of drive	▪ Experimental modal analysis to find shapes and frequency of vibrations (by strain gauges) ▪ Overall condition by vibrodiagnostic applied for gearbox, driveshaft bearings, particularly bearings at the pinion. ▪ Look for GMF (gear mesh frequency) & Natural frequency of large gears, their amplitude, sidebands and their spacing and amplitude. Also harmonics and sidebands at running speed of gears. Phase differences of 90 or 180° at bolts, base plates or foot of the components. All these items may help detection of wearing of teeth, misalignment, broken teeth, problems with backlash etc. For more details see Table 5.4. (gears related symptoms) ▪ SPM or similar technique appropriate for roller bearings of low speeds or displacements at plain bear-

(continued)

Table 5.1 (continued)

		ings.
CRUSHERS	Overall checks	▪ Crushing loads in operation by measuring deflection of supporting structure in 3 directions and use of FEM to calculate real loads in exploitation and control highly-stressed areas
	Foundations check (condition of concrete, supporting structure, levels disorder)	▪ Foundations sinking (deterioration); causes ▪ Concrete quality ▪ Oil or other fluids effects over foundation ▪ Cracks in metal construction, anchoring bolts corrosion of drive station foundations, base-plates ▪ NDT testing at carrying structure
	Condition of crusher body (shell)	▪ Visual control of loosen bolts, cracks, internal liners
	Condition of crusher rotors, hammers etc	▪ Visual control of pulleys, rolls and discs against cracks ▪ Condition of hammers ▪ Review history records of any repairs/replacement of wearing components
	Condition of crusher bearings	▪ Visual and control of wear at bearing, lubrication systems
	Condition of drives	▪ Control of drive shafts: internal defects or cracks in transition areas by NDT ▪ Review history records of any repairs ▪ Control of bearings against wearing, lubrication mal

(continued)

Table 5.1 (continued)

		functioning ▪ Control of couplings
	Overall operational condition by vibrodiagnostics	▪ Overall condition by vibrodiagnostic applied for gearbox, drive shaft & bearings, particularly bearings at the crusher rotors. ▪ Look for GMF (gear mesh frequency), rotors operational frequency, their amplitude, sidebands and their spacing and amplitude. Also harmonics and sidebands at running speed of gears. Phase differences of 90 or 180° at bolts, base plates or foot of the components. All these items may help detection of wearing of teeth, misalignment, broken teeth, problems with backlash etc. For more details see Table 5.4. (gears symptoms) ▪ SPM or similar technique appropriate for roller bearings of low speeds or displacements at plain bearings.
CRANES	Overall checks	▪ Standard load tests to confirm integrity and bearing capacity of structure
	Check of rails supporting structure (concrete or metallic) against level-disorder, overall integrity	▪ Geometrical measurements of any deflections (horizontal and vertical) of concrete and/or metal construction ▪ Check of dilatations ▪ Review historical records of any changes over time (cracks history-if any, former repairs, replacements of rails etc) ▪ Control of concrete struc-

(continued)

Table 5.1 (continued)

		ture, in particular any deflections of columns under influence of cranes operation ▪ Foundations sinking (deterioration); causes ▪ Concrete quality ▪ Oil or other fluids effects over foundation ▪ Cracks in metal construction, anchoring bolts and armature corrosion
	Check of rails (integrity, joints, deflections, cracks, wearing)	▪ Control of overall integrity of rails, joints and connections to base-plates ▪ Geometrical measurements of any deflections (horizontal and vertical) ▪ Check bolts and tightness ▪ Check expansion joints ▪ Visual control of carrying surface of rails and joints (uniformity of wearing marks) ▪ Review historical records of any changes over time (cracks history-if any, former repairs, replacements of rails etc)
	Check of load carrying structure	▪ Control of operational dynamical loads by strain measurements during operation in various operation modes of cranes ▪ Check of deflections regarding allowed values ▪ NDT control of structure
	Check of operators cabins	▪ Integrity of cabins ▪ Visual and NDT control of connections of cabins to cranes constructions ▪ Vibration measurements of cabins compared to al

(continued)

Table 5.1 (continued)

		lowed limits form H&S aspect of operators
	Check of drives	▪ Control of operational dynamical loads by strain measurements during operation ▪ Check the wearing marks on the wheels and rails ▪ Check overall condition of couplings, bearings (visual, clearances) ▪ Overall condition by vibrodiagnostic applied for gearbox, drive shaft & bearings
LOW-SPEED LARGE FANS	Foundations check (condition of concrete, supporting structure, levels disorder, anti-vibration pads)	▪ Foundations sinking (deterioration) or cracks ▪ Concrete quality ▪ Oil or other fluids effects over foundation ▪ Cracks in metal construction, anchoring bolts corrosion, baseplates ▪ NDT testing of supporting construction (at large fans only) ▪ Visual control of loosen bolts, cracks ▪ Control of functionality of stiffness of anti-vibration pads
	Impellers (cracks, wearing, marks from gasses flow, imbalance, wobbling)	▪ Visual control of overall condition (wearing, marks, cracks) ▪ NDT at weak points (front plates, connections to shafts, at corners of impeller blades) ▪ NDT of shafts ▪ Axial and radial wobbling (run-out) control

(continued)

Table 5.1 (continued)

	Connecting ducts, compensators, joints, housing of fan	▪ Control of metal-to-metal contacts ▪ Check of axial and radial clearances in cold and operating (hot) condition ▪ Check of false air intake or leakage ▪ Visual control of housing of fan ▪ Control tightens of bolts
	Overall vibrations at operating condition (including bearings) and temperatures	▪ Overall vibrations check, including run-up coast-down measurements. See Table 5.4 for more details controls ▪ Control of plain bearings by displacement (Orbit analysis) ▪ Control of roller bearings by accelerometers, SPM etc. ▪ Control of temperatures (level and change regimes) of bearings

Very good practice is to obtain quality assurance tests result from the suppliers, especially, NDT testing of structures, dimensional controls, balance protocols, etc.

Still, at the end is to the end user to perform some checks that are of major importance for the machinery as vibration baseline measurements, dimensional control, amps of main drive under load, etc. Good practices for technical assessment of new equipment are given in Table 5.2.

5.3 Typical Condition Monitoring Techniques

In order to select correct monitoring techniques, the typical characteristics and possibilities for application at low-speed machines are described for the most widely used techniques as follows.

5.3.1 Basics Tools L-L-T-S (Look–Listen–Touch–Smell)

"Look"—Visual inspection practices are the oldest and most common CM techniques employed in Industry that helps identify a broad range of potential problems

Table 5.2 Technical assessment practices for new equipment

Equipment	Condition assessment	Items to be checked
BALL MILLS	Foundations check (condition of concrete, supporting structure, levels disorder)	▪ Foundations correctness according documentation ▪ Concrete quality
	Condition of mill body (shell, trunions)	▪ Existence of cracks, voids, laminarities, or other internal defects in the body of the mill, particularly at the transition areas in trunions and around mill manholes by NDT (Dye penetrant, magnetic particles testing, Ultrasonic tests)
	Condition of trunion bearings	▪ Scratches or wearing at bearings, functionality of lubrication and other hydraulic systems
	Condition of drive-train	▪ Open gear transmission: damages at surface (Dye Penetrant or visual), ▪ Toothing at girth gears and pinions: backlash and top clearance, radial and axial wobbling, clearances at joints, tightening of connecting bolts, assembly of

(continued)

Table 5.2 (continued)

		the gears to the shafts or mill's body ▪ Condition of lubricant and functionality of lubrication system ▪ Bearings at drive shaft: (temperature) ▪ Gearbox (regular checks as for any gearbox: gears, bearings, lubrication)
	Overall operational condition by vibrodiagnostics	▪ Overall condition by vibrodiagnostic applied for gearbox, driveshaft bearings, particularly bearings at the pinion. Make a baseline measurements including GMF (gear mesh frequency) ▪ SPM or similar technique appropriate for roller bear ings of low speeds or dis placements at plain bear ings
ROTARY KILNS	Foundations check (condition of concrete, supporting structure, levels disorder)	▪ Foundations correctness according to documentation ▪ Concrete quality
	Condition of kiln shell & tyres	▪ Cracks, voids, laminarities, starting thickness or other internal defects in the kiln shell and tyres by NDT ▪ Visual control of tyres against wearing marks, scratches, pitting, spalling, other for surface damages, hot spot or even cracks ▪ Dimensional control of tyres against cone shape of tyres ▪ Visual control against any looseness or cracks in fixation systems of tyres

(continued)

Table 5.2 (continued)

	Condition of kiln axis and deformations	▪ Geometric measurements in cold or hot condition for axis condition (vertical and horizontal deflections) (to be related to foundations conditions also)
	Condition of supporting rollers & thrust rollers	▪ Rollers shafts: internal defects or cracks in transition areas by NDT ▪ Rollers body (carrying surface): *Visual* control of overall integrity, control against wearing marks, scratches, pitting, spalling for surface damages or even cracks; *Dimensional* control against cone shape of roller and comparison of dimensions of rollers at same supporting station ▪ Control of hydraulic system against functionality and control of axial travelling
	Condition of drive-train	▪ Open gear transmission: cracks at surface (Dye Penetrant), near under surface (Magnetic Particles effects), deeper in teeth or gear rim (US); pitting and other defects. ▪ Toothing at girth gears and pinions: backlash and top clearance, radial and axial wobbling, clearances at joints, tightening of connecting bolts, assembly of the gears to the shafts or mill's body

(continued)

Table 5.2 (continued)

	condition by vibrodiagnostics of drive	▪ Gearbox (regular checks as for any gearbox: gears, bearings, lubrication)
	Overall operational	▪ Overall condition by vibrodiagnostic applied for gearbox, driveshaft bearings, particularly bearings at the pinion including GMF (gear mesh frequency) of large gears, ▪ SPM or similar technique appropriate for roller bearings of low speeds or displacements at plain bearings
DRYERS	Foundations check (condition of concrete, supporting structure, levels disorder)	
	At rotary dryers:	
	o Condition of shell & tyres	The same as for rotary kilns
	o Condition of axis and deformations	
	o Condition of supporting rollers & bearings	
	Condition of drive-train	
	Overall operational condition by vibrodiagnostics of drive	
EXCAVATORS	Inspection of overall integrity	▪ Visual and functional controls
	Inspection of load car-	▪ Visual and NDT testing

(continued)

Table 5.2 (continued)

	rying structure	
	Condition of drives	▪ Overall condition by vibrodiagnostic applied for gearbox higher speed stages ▪ Visual
	Overall operational condition by vibrodiagnostics of drive	▪ Overall condition by vibrodiagnostic applied for gearbox, driveshaft bearings, particularly bearings at the pinion. ▪ SPM or similar technique appropriate for roller bear ings of low speeds or dis placements at plain bear ings.
CRUSHERS	Overall checks	▪ Visual and functionality
	Foundations check (condition of concrete, supporting structure, levels disorder)	▪ Foundations correctness according documentation ▪ Concrete quality ▪ NDT testing at carrying structure
	Condition of crusher body (shell)	▪ Visual control of loosen bolts, cracks, internal liners
	Condition of crusher rotors, hammers etc	▪ Visual control of pulleys, rolls and discs against cracks ▪ Condition of hammers
	Condition of drives	▪ Control of drive shafts: internal defects or cracks in transition areas by NDT ▪ Control of couplings

(continued)

Table 5.2 (continued)

	Overall operational condition by vibrodiagnostics	▪ Overall condition by vibrodiagnostic applied for gearbox, drive shaft & bearings, particularly bearings at the crusher rotors. ▪ SPM or similar technique appropriate for roller bearings of low speeds or displacements at plain bearings.
CRANES	Overall checks	▪ Standard load tests to confirm integrity and bearing capacity of structure
	Check of rails supporting structure (concrete or metallic) against level-disorder, overall integrity	▪ Geometrical measurements of any deflections (horizontal and vertical) of concrete and/or metal construction ▪ Check of dillatations. ▪ Concrete quality
	Check of rails (integrity, joints, deflections, cracks, wearing)	▪ Control of overall integrity of rails, joints and connections to base-plates ▪ Geometrical measurements of any deflections (horizontal and vertical) ▪ Check bolts and tightness ▪ Check expansion joints ▪ Visual control of carrying surface of rails and joints (uniformity of wearing marks)
	Check of load carrying structure	▪ Check of deflections regarding allowed values ▪ NDT control of structure
	Check of operators cabins	▪ Integrity of cabins ▪ Visual and NDT control of connections of cabins to cranes constructions ▪ Vibration measurements of cabins compared to allowed limits form H&S as

(continued)

Table 5.2 (continued)

		pect of operators
	Check of drives	▪ Check overall condition of couplings, bearings (visual, clearances) ▪ Overall condition by vibrodiagnostic applied for gearbox, drive shaft & bearings
LOW-SPEED LARGE FANS	Foundations check (condition of concrete, supporting structure, levels disorder, anti-vibration pads)	▪ Foundations correctness according documentation ▪ Concrete quality ▪ Cracks in metal construction, anchoring bolts corrosion, baseplates ▪ NDT testing of supporting construction (at large fans only) ▪ Visual control of loosen bolts, cracks ▪ Control of functionality of stiffness of anti-vibration pads
	Impellers (cracks, imbalance, wobbling)	▪ Visual control of overall condition (wearing, marks, cracks) ▪ NDT at weak points (front plates, connections to shafts, at corners of impeller blades) ▪ NDT of shafts ▪ Axial and radial wobbling (run-out) control
	Connecting ducts, compensators, joints, housing of fan	▪ Control of metal-to-metal contacts ▪ Check of axial and radial clearances in cold and operating (hot) condition ▪ Visual control of housing of fan ▪ Control tightens of bolts

(continued)

Table 5.2 (continued)

	Overall vibrations at operating condition (including bearings) and temperatures	▪ Overall vibrations check, including run-up coast-down measurements. ▪ Control of bearings by accelerometers, SPM etc. ▪ Control of temperatures (level and change regimes) of bearings

like loose or worn parts; leaks of oils or any liquids; missing parts; poor electrical, or pipe connections; etc. Also, human sensory-based inspections can verify the results from other CM techniques. However, there can be one drawback that in some cases observers can become so familiar with their surroundings that changes of interest go unnoticed.

"Listen"—Audio inspection practices help identify a broad range of potential problems, including worn high-friction bearings, steam leaks, pressure relief valve leaks or discharges, coupling leaks, excessive loading on pumps, poor mechanical equipment alignment, etc. Humans are particularly sensitive to be new or changed sounds and are easily taught to report and investigate unusual sounds. Experience is required to distinguish unusual sounds from regular ones.

"Touch" as an inspection technique can be also extremely useful as it can detect heat, scaling, and roughness changes. Human touch is extremely sensitive and able to differentiate surface finish differences not discernable by eye. However, the inspectors are to be very carefully with this technique because a person may be burned or injured by touching hot, moving, or under voltage operating equipment.

"Smell" is an inspection technique that goes together with the previous techniques. Potential problems as oil spillages, leakage of gasses (e.g., propane) or vapors, burned rubber parts (e.g., seals) can be detected which require further additional inspections.

Although these tools seem simple, still they can support proactive behavior. It is on the user of these tools to differentiate early what is deviating from regular, why there is some knocking in the machine, or why a simple touch is showing slight difference in the vibrations or strokes in the gear, etc. and undertake some actions to see why is like that and if required undertake early corrections.

5.3.2 Vibrodiagnostics

Vibrodiagnostics as inspection technique is one of the most widely used techniques. It is based on the energy emitted from the measured equipment while being in operation. Using different analyzing techniques, different types of errors in the equipment can be recognized as this energy varies in quantity and quality properties. It is applicable for all moving parts, rotating, or reciprocating mechanisms.

The most commonly used methods for vibrations analysis are the overall descriptors (as RMS or Max values) and analysis in Time domain or Frequency domain (FFT, Power spectra analysis, etc.). However, no matter of what method is used, three different parameters may be employed: displacement, vibration, or acceleration. For low-speed machines, depending on the actual speed, displacement, or velocity it can be used. Although still major measuring parameter is the displacement of separate components, nowadays, with special probes, some measurements are performed using the pulse energy of the shocks thus detecting developing faults at very accurate way.

Actually, the proactive CM uses the **vibration condition monitoring** as a development base. The particularity is that for low-speed machines linear displacement is taken as measuring quantity based on the anticipated frequency of excitation [30] following ISO13373-1:2002. Considering machines complexity, following measurements can be employed:

- vibration measurements expressed as r.m.s displacement;
- vibratory displacement (relative) as S_{max} (zero-to-peak) and S_{pk-pk} (peak-to-peak) as low-speed machines often show predominantly sinusoidal vibrations [31].

A part of these direct parameters, more **complex analysis** can be used involving Fast Fourier Transform (FFT) for recognition and separation of the faults related to different components of the machine.

Considering the low-speed of machines in interest, **the measurement quantities** should be focused on low frequency areas. Actually, the frequency range should be tailored to the specific machine. Noncontacting probes are to be used whose output is directly proportional to the relative displacement between the rotating and non-rotating elements of the machine. This is also recommended by ISO13373-1:2002 as shown on Fig. 5.1.

But, apart of these analyses, for low-speed machines very beneficial information about the machinery health can be obtained using specialized methods as Shock pulse measurements and Envelope analysis. In the last few years, the energy of the shock pulses is used for detection of damages not only in rolling bearings, but also in gears and similar applications as the wide dynamic range of these piezo-ceramic transducers provide accurate detection of the early damages in metallic components coming into contact when still sharp edges exist generating strong pulse energy.

Such **measurement systems** should be connected and incorporated in the controlling systems of the process in order to have complete online monitoring and possibility for reactions in real time. Therefore, all equipment should be commercially available at low costs for the plant. Taking reference measurements and creating a **baseline** is also of major importance and it consists of creating group of descriptors such as S_{pk-pk}, S_{max}, FFT graphs, shape of filtered curves describing vibratory motion of some component or shape of wearing surface under different, predefined process states.

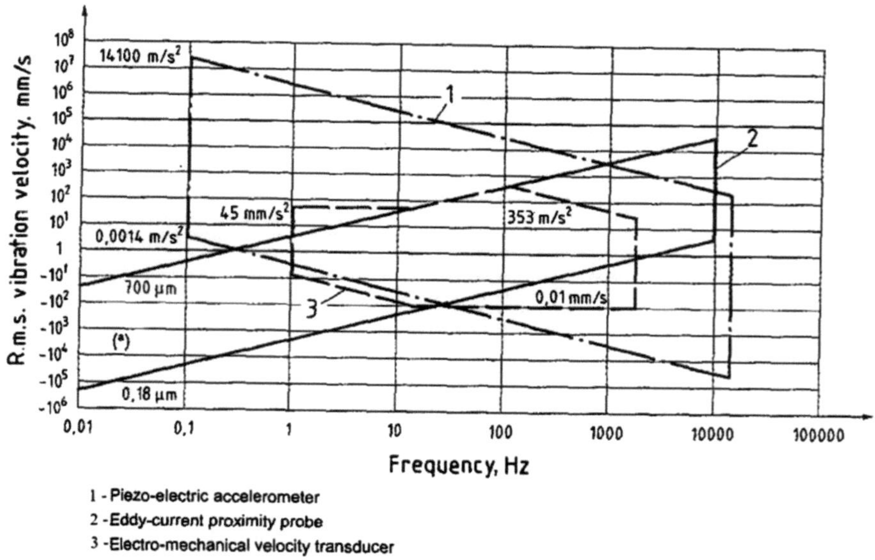

Fig. 5.1 Dynamic range versus frequency range for the application of transducers typically used for machine condition monitoring [30]

In order to point out differences when aiming to proactive information, regarding the standard vibration condition monitoring procedure, flowchart according to ISO13373-1 is given with marked changes on Fig. 5.2.

For purposes of proactive CM as a separate **package of diagnostics rules should be set** based on the well-known vibration quantities, FFT analysis, and additional tailored tools created on mathematical base for extraction of phenomena or shapes of interest for the particular machine. They are to be linked directly to physical phenomena and processes in the machine thus providing simple rules for conclusions related to the machines condition that are to be applied by operators or maintenance staff.

Typical applications of the various analysis techniques used for low-speed machines are shown in Table 5.3 [9, 17, 18, 21, 24].

In relation to time signals, except the visual presentation of how the vibrations are changing in time and analysis of the periods of repetitive spikes, this technique is also related to the statistical methods for investigation of random signals. The simplest method is to use overall root-mean-square (RMS) level and crest factor, i.e., the ratio of peak value to RMS for detection of localized defects. Probability density has also been used popularly for bearing defect detection and statistical moments, the skewness as measure for the relative energy above and below the mean level and kurtosis—a measurement between the insensitive lower moments and the over-sensitive higher moments being useful in the detection of bearing failure.

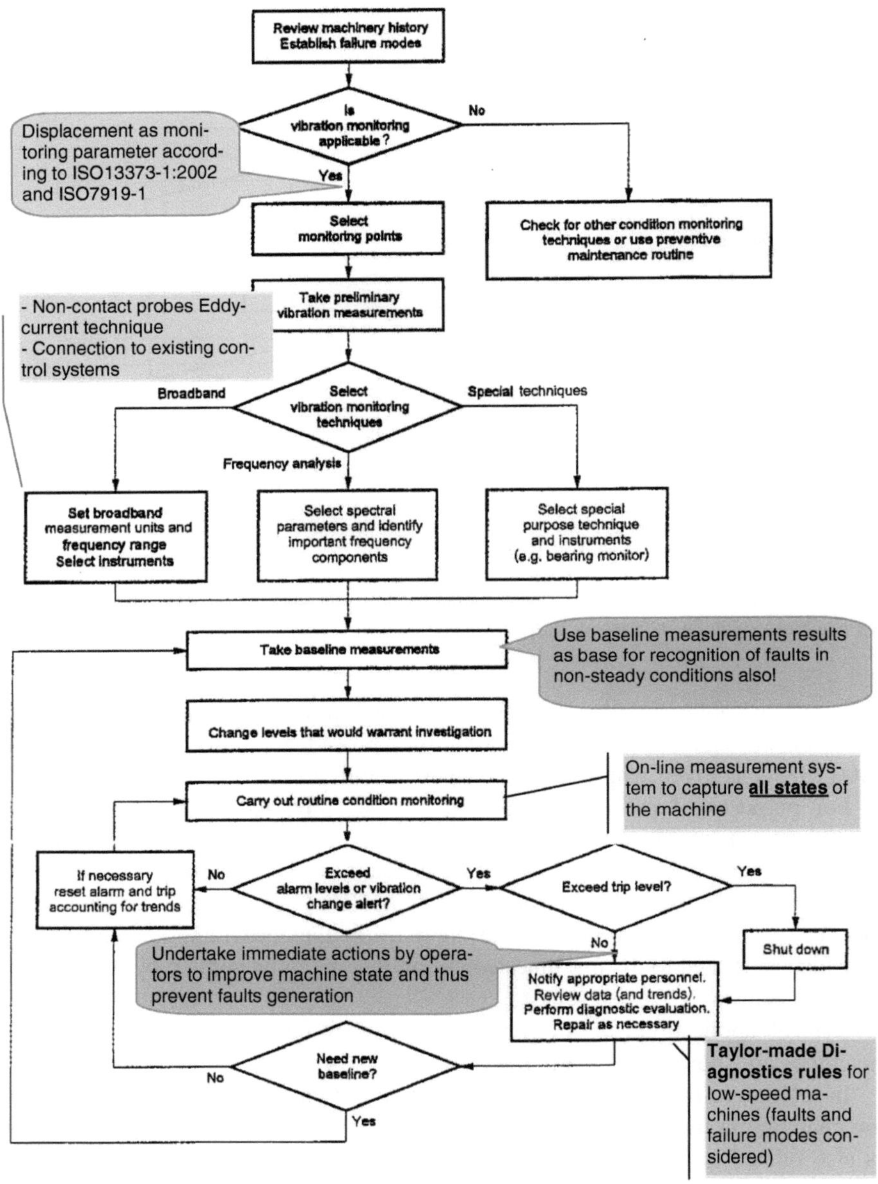

Fig. 5.2 Vibration monitoring flowchart with linear displacement as measuring quantity for proactive CM

The meaning of these parameters is the following:

RMS: is a statistical measure of the magnitude of a varying quantity and in terms of vibration, and it is indication for the overall vibrartion energy. The "RMS" value is calculated by simply multiplying the peak amplitude (shown in the graphic

Table 5.3 Review of applicable techniques for vibration monitoring [9, 17, 18, 21, 24]

Technique	Application	Fault/machine
Time signal	Waveform visualization for identification of distortion	Rubbing, impacts, clipping, cracked teeth, general presentation of periodicity in machinery movement
Frequency analysis	Identification and analysis of frequency components in complex signals	Out-of-balance, misalignment, defects at gears, roller and plain bearing, bended shafts, couplings etc. at any rotating or reciprocating machine
Phase	Operational deflection shapes	Detection of developing cracks in shafts. Used for balancing
Shock pulse method	Detection of damages or faults in lubrication and early damages of bearings and gears	Defects at very early stages or lubrication faults at roller bearings, defects at separate components of bearings. Applicable for low-speed roller bearings and gears
Cepstrum	Identification and separation of families of harmonics	Rolling elements bearing, bladed machines, gearboxes
	Identification and separation of families of sidebands	
Envelope analysis	Amplitude demodulation Observation of a low-frequency amplitude modulation at high frequency	Rolling element bearing, electrical machines, gearboxes
Dynamic crest	Calculation of high-pass filtered factor signals	Faults in low-speed machines
Synchronous time	Improving signal-to-noise ratio averaging Waveform analysis Separating effects of adjacent machines	Electrical machines, reciprocating machines, gearboxes, etc.
	Separating effects of different shafts	
Impact testing	Resonance testing	Foundations, bearings, couplings, gears
Scan analysis	Analysis of nonstationary signals	Fast run-up/coast down

below) by 0.707. In the case of a set of n values $\{x_1, x_2, \ldots, x_n\}$, the RMS can be calculated as:

$$x_{\mathrm{rms}} = \sqrt{\frac{1}{n}\left(x_1^2 + x_2^2 + \cdots + x_n^2\right)} \tag{5.1}$$

Although it gives good rough indication of the vibration severity, this parameter should be used with caution for low speed machines due to their low vibration impact energy (see the example in Chap. 3). Relatively low RMS value can still mean big deviations in machine components movement, so it is highly recommended to use this parameter as a trend indication (to follow its change over time and not as onset parameter for condition assessment).

Crest factor is a measure of a waveform, showing the ratio of peak values to the average value (peak amplitude of the waveform divided by the RMS value of the waveform):

$$C = \frac{|x|_{\text{peak}}}{x_{\text{rms}}} \tag{5.2}$$

where:

- $|x|_{\text{peak}}$ is peak amplitude
- x_{rms} is RMS value.

In other words, crest factor indicates how extreme the peaks are in a waveform. It is a quick and useful calculation that gives an idea of how much impact is occurring in a time wave. This is an useful information that is lost if one is only viewing a spectrum as the FFT cannot differentiate between impacting and random noise. Impacting in a time waveform may indicate rolling element bearing wear, gear tooth wear, or cavitation. As for other machinery, also for low-speed machinery, the crest factor should be trended over time in order to see if the amount of impacting is increasing or not, just that the absolute values will be again of lesser amount comparing to higher speed machines.

Skewness is a measure of the asymmetry of the probability distribution of a real-valued random variable about its mean. The skewness value can be positive or negative, or even undefined (See Fig. 5.3).

For a sample of n values, a natural method of moments estimator of the population skewness is:

$$b_1 = \frac{m_3}{s^3} = \frac{\frac{1}{n}\sum_{i=1}^{n}(x_i - \bar{x})^3}{\left[\frac{1}{n-1}\sum_{i=1}^{n}(x_i - \bar{x})^2\right]^{3/2}}, \tag{5.3}$$

where:

- $\bar{x}$ is the sample mean,
- s is the sample standard deviation, and
- the numerator m_3 is the sample third central moment.

Skewness is actually an indicator of the symmetry of the signal around the mean value. In terms of vibrations, in good condition of the machinery, it should be around zero. The higher asymmetry, the higher irregularities in the time signal or presence of some impacts or other irregularities in the machine. Considering it takes the power of 3 of measured values, it "exaggerates" the high values and suppresses the low values, so it is more sensitive parameter to the asymmetry in high values than the simple mean values thus making it more useful for low-speed machinery

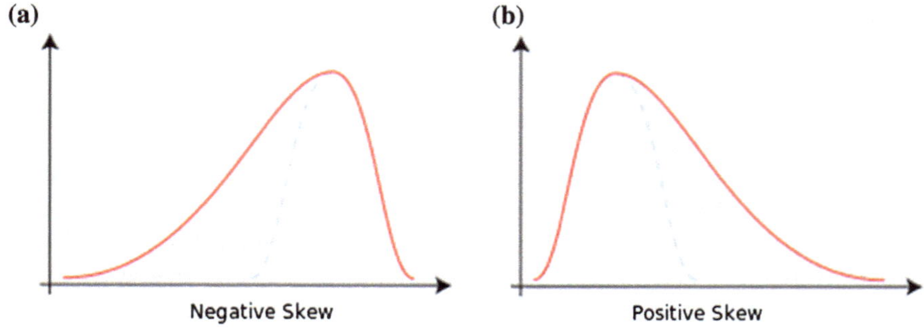

Fig. 5.3 Typical skewness [32]. **a** Negative skew. **b** Positive skew

monitoring, although it is not widely used in industrial practices it requires more knowledge about its meaning and analysts getting familiar to the values.

Kurtosis is a measure of the "peakedness" of the probability distribution of a random variable. In a similar way to the concept of skewness, kurtosis is a descriptor of the shape of a probability distribution (see Fig. 5.4).

For Gauss distribution, kurtosis has value of 3, so in the industrial application, more often an excess kurtosis is used by simple subtracting the value 3 of the total estimated value.

For a sample of n values the sample excess kurtosis is:

$$g_2 = \frac{m_4}{m_2^2} - 3 = \frac{\frac{1}{n}\sum_{i=1}^{n}(x_i - \bar{x})^4}{\left(\frac{1}{n}\sum_{i=1}^{n}(x_i - \bar{x})^2\right)^2} - 3$$

where:

- $\bar{x}$ is the sample mean,
- m_4 is the fourth sample moment about the mean,
- m_2 is the second sample moment about the mean.

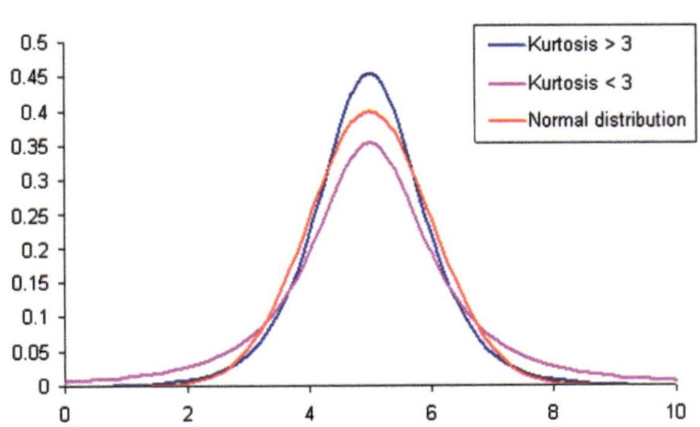

Fig. 5.4 Typical kurtosis values [32]

A set of data with a high kurtosis value will produce a distribution curve with higher peak value at the mean and longer "tails", or in other words, more data points at the extreme values from the mean.

Kurtosis has significant application in AEs for recognition of damaged bearings (healthy bearing has Kurtosis of 3), but it has to be used with certain precautions as with further development of bearing damage, the kurtosis value is reversed again to close to 3, so trending is the most important in this case too.

However, still the most commonly used is the **FFT analysis**. The FFT takes the complex waveform and breaks it down into the component sine waves. The amplitudes for each sine wave is then plotted at the frequency of the sine wave, creating the Spectrum (see Fig. 5.5).

One of the best summarized overviews for FFT applications to recognize failures were given by [3] and based on that, set of major faults applicable not only for low-speed machines, but also for medium-and high-speed machines with their symptoms is presented in Table 5.4.

When using vibrodiagnostic techniques, apart from discovering faults developing in the machines, these techniques in proactive CM should be used in a way to detect phenomena that although not critical for machinery health itself, they can bring to development of other more serious faults and failures. For example, a bent shaft or misalignment can be not so critical for machinery health and left uncorrected in all other cases, but for proactive purposes this phenomena should be linked to what consequences can it make, e.g., there can be increased load or wearing in a bearing, increased vibrations that affects rest of the components in the machine. Knowing if the secondary effects can be critical, certain corrections should be undertaken on time. Even more, with the proactive CM, an analysis should be made and discover why bent shaft or misalignment appeared at first place (if the installation was in accordance to valid limits)? Maybe something has caused loose of bolts, deflection of structures, etc.

All of the above given applications for FFT are valid for low speed with one recommendation that attention should be paid on the frequencies of interest. Low rotational speed will ask for analysis of lower range of frequency domain, while

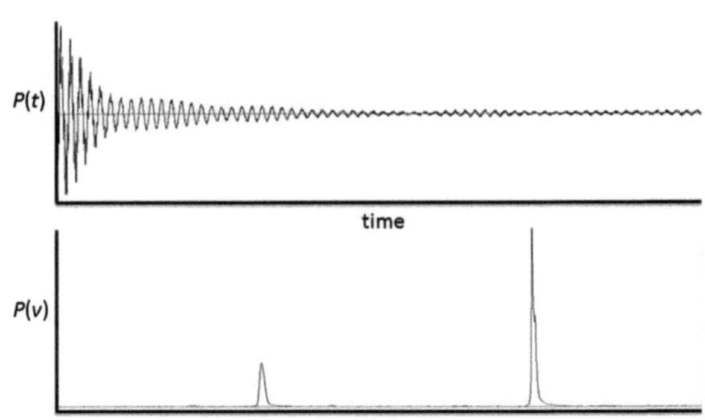

Fig. 5.5 Conversion of time signal to frequency spectrum [32]

Table 5.4 Possible faults to symptoms relationships [3]

Fault	Symptoms
Bend Shaft	High axial vibrations with axial phase differences tending towards 180° on same machine component. Dominant vibration normally occurs at 1× if bent near shaft centre, but at 2× if bent near coupling
Misalignment	
Angular	High axial vibrations, phase difference 180° across the coupling. Dominant vibration is usually axial at both 1 and 2× rpm, although it can be one dominant at 1, 2 or 3×. Severe misalignment can excite many 1× rpm harmonics, but without rised noises on the spectra
Parallel	Similar symptoms to angular, but high radial vibrations with phases difference near 180° across the coupling. 2× harmonic can be higher than 1×, but it depends on coupling design. Severe misalignment can cause higher peaks at higher frequencies (3, −8×), but typically without raised noise floor
Misaligned bearing cocked on shaft	Higher axial vibration. This can cause twisting motion and generation of phase difference about 180° top to bottom and/or side to side measured in axial direction on same bearing housing
Resonance	It is shown by a 90° phase change at the point when the forcing frequency coincides with a natural frequency, and approaches a full 180° phase change when the machine passes through the natural frequency (depending on the amount of damping present)
Mechanical looseness	Type A: caused by structural looseness/weakness of machine feet, baseplate or foundation, also by deteriorated grouting, loose hold down bolts at the base, distortion of the frame or the base (i.e. soft foot). Phase differences approx. 90–180° between vertical measurements on bolt, machine foot, baseplate or base itself
	Type B: caused by loose pillow block bolts, cracks in frame structure or in bearing pedestal
	Type C: unstable phase, especially from one to other measurement. Very different readings may occur in different measurements at 30° increments in radial direction around housing. Looseness will often cause sub-harmonics multiplies exactly at ½ or 1/3× rpm (0.5, 1.5, 2.5× etc.)
Journal bearings	
Wear/clearance problems	Presence of series of running speed harmonics (up to 10 or 20). Wiped journal bearings often allow high vertical amplitudes compared to horizontal, but may show pronounced peak at 1× rpm. Excessive clearance may allow minor unbalance and/or misalignment to cause high vibration than those at normal bearing clearance
Oil whirl instability	Occurs at 0.40–0.48 rpm and often very severe. It is considered as excessive when amplitude exceeds 40 % of bearing clearance. This is oil fill excited vibration where

(continued)

Table 5.4 (continued)

Fault	Symptoms
	deviations in normal operating conditions (attitude ration and eccentricity) cause oil wedge to "push" shaft around within bearing. Destabilizing force results in a whirl (forwards precession). Oil whirl is unstable since it increases centrifugal forces which increases whirl forces. Can cause oil to no longer support shaft and can become unstable when whirl frequency coincides with rotor natural frequency. Changes in oil viscosity, lube pressure and external preloads can affect oil whirl
Oil whip instability	Occurs when machine operates at or above 2× rotor critical frequency. When rotor brought up to twice critical speed, whirl will be very close to rotor critical and may cause excessive vibration causing that the oil is no longer capable of supporting. Whirl speed will actually "lock onto" rotor critical and thus peak will not pass through it even if machine is brought to higher and higher speeds. Produces a lateral forward processional subharmonic vibration at rotor critical frequency. Inherently unstable which can lead to catastrophic failures
Rolling element bearings (four failure stages)	Stage 1: earliest indications of bearing problems appear in ultrasonic frequencies from about 250,000–350,000 Hz; later, as wear increases, usually drops to approximately 20,000–60,000 Hz (1,200,000–3,600,000 CPM). These are frequencies evaluated by Spike Energy (gSE), HFD (g) and Shock Pulse (dB). Acquiring high frequency enveloped spectra confirms whether or not bearing is in starting failure
	Stage 2: slight bearing defects begin to "ring" bearing component natural frequencies which predominantly occur in 30–120 K CPM range. Such natural frequencies may also be resonances of bearing support structures. Sideband frequencies appear above and below natural frequency peak at end of Stage 2. Overall spike energy grows (for example, from 0.25 to 0.50 gSE)
	Stage 3: bearing defect frequencies and harmonics appear. When wear progresses, more defect frequency harmonics appear and number of sidebands grow, both around these and bearing component natural frequencies. Overall spike energy continues to increase (for example, from 0.5 to over 1 gSE). Wear is now usually visible and may extend throughout periphery of bearing, particularly when many sidebands accompany bearing defect frequency harmonics. High frequency demodulated and enveloped spectra help confirm Stage 3
	Stage 4: toward the end, amplitude of 1× rpm is affected. It grows and normally causes growth of many running speed harmonics. Discrete bearing defect and component natural frequencies actually begin to "disappear" and are replaced by random, broadband high frequency "noise floor". In addition,

(continued)

Table 5.4 (continued)

Fault	Symptoms
	amplitudes of both high frequency noise floor and spike energy may in fact decrease; but just prior to failure, spike energy, and HFD will usually grow to excessive amplitudes
Gears	
Normal spectrum	Normal spectrum shows Gear and Pinion speeds, Gear Meshing Frequency (GMF) and very small GMF harmonics. GMF harmonics will have usually running speed sidebands around them. All peaks are low amplitude and no natural frequencies of gears are excited
Tooth wear	Key indicator is excitation of Gear Natural Frequency (fn), sidebands around it spaced at the running speed of damaged gear. GMF may or may not change in amplitude, but amplitudes and number of sidebands increases around GMF when wear is noticeable. So, sidebands are better indicators of wear than GMF. Higher amplitude commonly occurs at either 2× or especially 3× GMF
Tooth load	GMF are often very sensitive to gear loads and especially if sidebands remain low and no gear natural frequency are excited, it means there is no problem. From this reason, measurements should take place at maximum operating load for meaningful spectra comparison
Gear eccentricity, backlash or non-parallel shafts	Fairly higher amplitude sidebands around GMF harmonics because the gear eccentricity, higher backlash and non-parallel shafts allow the rotation of one gear to "modulate" either the GMF amplitude or the running speed of the other gear. The gear with problem is indicated by the space of the sidebands frequencies. 1× rpm level of eccentric gear will be normally high if eccentricity is the dominant problem. Improper backlash normally excites GMF harmonics and Gear Natural Frequency, both of which will be sidebanded at 1× rpm. GMF amplitude will often decrease with increasing load if the backlash is the problem
Misalignment	It excites second order or higher GMF harmonics which are sidebanded at running speed. Often will show only small amplitude at 1× GMF, but much higher levels at 2 or 3× GMF harmonics. Also, sidebands around 2× GMF will often be spaced at 2× rpm. Often sideband amplitudes can be not equal on left and right side of GMF and GMF harmonics due to tooth misalignment. It causes uneven wear pattern
Cracked/broken tooth	It will generate high amplitudes at 1× rpm of this gear only in the TIMEWAVEFORM, and it will excite gear natural frequency (fn) sidebanded at its running speed. It is best detected in Time Waveform which will show pronounced spike every time the problem tooth raised to mesh with teeth on the matting gear. Time between impacts Δt will corresponds to 1/rpm of gear with the problem. Amplitudes of

(continued)

Table 5.4 (continued)

Fault	Symptoms
	impact Spikes in Time Waveform will often be 10 or 20× higher than that at 1× rpm in the FFT
Gear assembly phase problem	Gear Assembly Phase Frequency (GAPF) can result in Fractional Gear Mesh Frequencies. It literary means (TG/NA) gear teeth will contact (TP/NA) pinion teeth and will generate NA wear patterns, where NA in a given tooth combination equals the products of prime factors common to the number of teeth on the gear and pinion (NA—Assembly Phase factor). GAPF or harmonics can show up right from the beginning if there were manufacturing problems. Also, its sudden appearance in a periodic survey spectrum can indicate damage to the teeth in mesh at the time of ingestion just as they enter and leave meshing or that gears have been reoriented
Hunting tooth problems	Hunting tooth frequency (fHT) occurs when faults are present on both: girth gear and pinion which may have occurred during the manufacturing process, due to misalignment, or in the field. It's usually at low frequencies (<600 CPM), it is often missed. A gear set with this tooth repeat problem normally emits "growing" sound from the drive. The maximum effect occurs when the faulty pinion and gear teeth both enter mesh at the same time (on some drives this may occur only one at every 10–20 revolutions)
Loose bearing fit	Excessive clearance of bearing supporting the gears can not only excite many running speed harmonics, but will often cause high amplitude response at GMF and/or 2 and 3 GMF. These high GMF amplitudes are actually a response to, and not the cause of looseness within the bearings supporting the gearing. Such excessive clearance can be caused either by extensive bearing wear or by improper bearing fit onto the journal during installation. Left uncorrected, it can cause excessive gear wear and damage to the other components
Beat vibrations	A Beat Frequency is the result of two closely spaced frequencies going into and out of synchronization with one another. The wideband spectrum normally will show one peak pulsating up and down. When you zoom into this peak (lower spectrum bellow), it actually shows two closely spaced peaks. The difference in these two peaks (F2–F1) is the beat frequency which appears itself in the wideband spectrum. The beat frequency is not commonly seen in normal frequency range measurements since it is inherently low frequency, usually ranging from only approximately 5–100 CPM
	Maximum vibrations will result when the time waveform on one frequency (F1) comes into phase with the waveform of the other frequency (F2). Minimum vibration occurs when waveforms of these two frequencies line up 180° out of phase
Soft foot, sprung foot and foot-related resonance	"Soft foot" occurs when machine foot or frame deflects greatly when a hold-down bolt is loosened to hand tightness, causing the foot to rise more than approximately 0.05–0.08 mm. This

(continued)

Table 5.4 (continued)

Fault	Symptoms
	does not always cause a great vibration increase. However, it can do so if the soft foot affects alignment of motor and gap concentricity
	"Sprung foot" can cause great frame distortion, resulting in increased vibration, force and stress in the frame, bearing housing etc. This can occur when a hold-down bolt is forceably torqued down on the sprung foot in an attempt to level the foot
	"Foot related resonance" can cause dramatic amplitude increase from 5 to 15× or more, as compared with that when the bolt (or combination of bolts) is loosened to hand tightness. When tight, this bolt can notably change the natural frequency of the foot or machine frame itself
	Soft foot, Sprung foot or Foot-related Resonance most often affects vibration at 1 rpm, but can also do so at 2, 3× rpm, 2× line frequency, blade pass frequency etc.(particularly Foot-related Resonance)

bearings will depend of the actual shaft speeds. Lately recommended procedure is to create patterns on the specific machines prior doing analysis that would appoint all important frequencies thus analyst could look only to these ranges of interest and note any changes.

5.3.3 Ultrasonic Technique

Ultrasound is defined as sound waves that have frequency levels above 20 kHz (the human range is commonly between 20 and 20,000 Hz). Most machines emit steady sonic signatures under normal operating conditions and any changes in these signatures can point to start of wearing or deterioration of some components. This is actually what promotes application of this technique for identification of bearing damages, compressed air, or hydraulic fluid leaks, vacuum leaks, etc. [33].

The AE technique also deals with signals in the high-frequency range and has been increasingly used for condition monitoring of rotating machinery as well as structures. However, the AE technique differs from the ultrasonic technique in terms of the frequency range of interest and parameters for condition assessment. The AE technique generally operates in the 100 kHz–1 MHz while the ultrasonic technique focuses on the frequency range of 20–100 kHz. Parameters such as ring down counts, events, rising time, duration, and peak amplitude are normally used in the AE technique to examine abnormality [34]. In the ultrasonic detector, abnormality is usually detected by listening to the characteristics of sound or RMS indicator on the panel.

The heterodyne circuit is the main component in ultrasound detectors as it takes the ultrasound signal detected by the transducer, and converts it into an audible signal or can be processed through a converter to obtain a quantitative output in decibels (dB).

The main advantage of ultrasound is the high signal-to-noise ratio detection ability providing quite exact localization of the energy source of the ultrasound activity, regardless of environmental interferences (i.e., noise). This technique and measuring systems are not so typical and frequently applied in industry, but with the time, their availability is becoming easier, thus, their application is also increasingly extended in the current condition monitoring systems.

5.3.4 Lubricants Condition Monitoring

Lubrication is one of the essential activities in maintenance. It is well known that equally important are the primer selection of the oil as function of the application, as well as follow-up of the changes in the lubricant properties as indicators for what is happening in the machine-time replacement. There are several analyses of the lubricants condition, which are very helpful in discovering early generation of faults in the lubricated machines. If paid proper attention and go into the root causes of any property change, it can serve as one of the most useful proactive condition monitoring tool. Their use is given as follows [33]:

Ferrography is a microscopic examination and analysis of debris (particles) found in lubricating oils. These particles consist of metallic and nonmetallic matter. The metallic particles may appoint to normal rubbing wear, cutting wear, severe sliding particles, bearing wear (fatigue spall particles, laminar particles), and gear wear (pitch line fatigue particles, scuffing, or scoring particles). Nonmetallic particle consists mainly of dirt, sand, or corroded metallic particle. Through identification of the density and size ratio of particles in oil or grease phenomena such as early and advanced wear, fatigue and/or corrosion can be detected.

Particle counter testing monitors particles in both lubricating and hydraulic oils caused by problems such as corrosion, wear, fatigue and contaminants, but without information about chemical nature of the particles. There are more types of particle counting tests available, but most commonly used are the light extinction and light scattering particle counters. A direct reading of the ISO cleanliness value can be determined from the both tests.

Sediment testing provides information about sediment at low-viscosity oils (e.g., inorganic sediment from contamination and organic sediment from oil deterioration or contamination) and soluble sludge from electrical insulating oil deterioration. It involves the use of a centrifuge to separate sediment from oil, and the sediment-free portion is subject to further steps to measure the soluble sludge. The total sediment is further processed to estimate organic/inorganic composition.

This technique can be used to analyze petroleum-based insulating oils in transformers, breakers, and cables.

Atomic emissions spectroscopy identifies problems such as corrosion, wear metals, contaminants, and additives in lubrication and hydraulic oil samples by measuring the characteristic radiation emitted when samples are subjected to high energy and temperature conditions. The test results show presence (in ppm) of iron, aluminum, chromium, copper, lead, tin, nickel, and silver, and components of oil additives such as boron, zinc, phosphorus, and calcium. This technique can be used to analyze oil used in diesel and gasoline engines, compressors, transmissions, gearboxes, and hydraulic systems.

Infrared spectroscopy is used to determine the level of an element in a sample without destroying the sample by measuring the absorbent light energy at various specific wavelengths. The analysis can provide information about oil deterioration, oxidation, water contamination, or oil additives. This technique can be applied to turbine generators, transformer oils, and breakers.

Potentiometric titration—Total Acid Number (TAN) is used to determine the extent of breakdown in lubrication or hydraulic oil by determining the level of acidity in an oil sample (limited to petroleum-based oils). The test involves mixing the oil sample with solvents and water and then measuring the change in the electrical conductivity. This technique can be used to test oil used in diesel/gasoline engines, gas turbines, transmissions, gearboxes, compressors, hydraulic systems, and transformers.

Total Base Number (TBN) measures the amount of active additive left in a sample of oil. Oils are continually exposed to acidic compounds which cause the oil to turn more acidic. This is particularly true of crankcase oils. In order to give longer life-time of oil-alkalinity, the manufacturers give the oil a 'reserve alkalinity' to fight any acidity which forms in the oil during use. A higher TBN means the oil has more reserve alkalinity available which can be used to reduce the corrosive effects of acids.

The **Karl Fischer titration** test measures moisture in a lubrication or hydraulic oil sample, which is an indicator of a degraded oil condition, by measuring electrical current flow between two electrodes immersed in the sample solution. Karl Fischer reagent is metered into the sample until all of the entrained water is reacted with the reagent. Results are reported in ppm of water. This technique can be used to analyze oil in enclosed oil systems such as engines, gearboxes, transmissions, compressors, hydraulic systems, turbines and transformers.

The **kinematic viscosity** test provides an indication of oil deterioration over time or contamination of the oil by fuel or other oils. The test measures the fluids resistance to flow under known pressure and temperature conditions and involves forcing a sample to flow through a capillary viscometer. Based on the test results, the dynamic viscosity of the oil sample can be calculated. This technique can be used to test oil used in diesel/gasoline engines, turbines, transmissions, gearboxes, compressors, and hydraulic systems.

Dielectric strength tests are used to measure the insulating quality of electrical insulating oils, which are often deteriorated by contamination or oil breakdown.

The test is performed by subjecting the sample to an electrical stress at a given temperature by passing voltage through the sample. This technique can be used to test insulating oils in transformers, breakers, and cables.

5.3.5 Geometric Measurement

Geometry measurements are not so widely used in maintenance terminology, but they are especially significant for low-speed machines and lately there is rising trend of using the photogrammetry measurements also for condition monitoring purposes in machinery. These measurements are used to measure any deterioration in the machine structure, fundaments, deformations of the rotary machines axis, deformations of machines structures, wearing of large trunnion bearings, misalignment of drive trains, change in locations of supports, and forces acting by checking the shapes and distances (dimensions). They are usually performed by laser instruments with sufficient accuracy at certain regular intervals or by use of better quality cameras to record previously placed markings on control points. It is very important to determine what parameters will be measured and set expected limits for change. They are strongly recommended at large supporting structures of rotary excavators, structures at large conveyors, grid constructions, crane galleries (both concrete and metal structure), large rotary body machines, supports of ducts with thermal loads etc.

When arranging geometrical measurements, the first thing to do is to create simplified model of the machine with all axes and supports and determine all movements of the component. Measurements should make a footprint of the machine condition comparing to its original assembly and focus on possible deviations that affect matching components. Therefore, following should be measured:

- **Position and possible misalignment of all axes of rotation** of the main bodies and moving components such as drive shafts, gears, rollers, etc. Position should be described in the two planes or in vertical, horizontal plane;
- **Foundations settlement**, or trough marked points measure their relative position comparing to surrounding structures or some critical items of the machine. For example, at ball mills, the sinking of one trunnion foundation may cause significant wearing of the other bearing and disturb the lubrication process if there will be hydrodynamic bearings.
- **Deformation (deflection) of main supporting structure** (main beams).

Within these measurements, recording movies of machines operations and using specialized software the machine condition can be estimated by focusing on previously defined lines and points for monitoring.

Another quite a new technique for geometric measurements is **photogrammetry method**. It bases on analyses of photographs taken on the tested object, with the use of high-quality cameras (optics) and in presence of scaling bars, crosses, and marker points. Based on such photographs specially developed software calculates geometric parameters. This method enables identification of 2D and 3D geometry with very high efficiency and accuracy.

5.3.6 Strain Measurements

Very often in low-speed machines forces can act outside the designed limits and cause additional stresses in the structures. The consequences are difficult to be recognized until a large deformation or cracks appear. Therefore, it has been proven as useful practice to perform strain measurements to find [35]:

- The stresses generated in the bodies or load carrying constructions while machines are in operation (operational loads stresses).
- Residual stresses in the structures; or
- Measure operational loads with previously calibrated strain gauges.

These measurements actually give clear picture of what is happening with machines at operation and predictions can be made on the risks of some faults generation. There are several methods for strain measurements and they can be grouped into two main groups determined by the application:

- Full-field methods (for macroscopic effects before looking into details) answering questions as where will stress be greatest and how extensive a region must be studied. It included several techniques:
  - Optical interference (Moire)
  - Holographic
  - Interferometric
  - Brittle coatings
  - Grids and rulings
  - Photoelastic methods.
- Point measurement (detailed study of limited region)
  - Bonded electrical resistance strain gauges
  - Extensiometers (clip gauges).

The most common measuring method for condition monitoring of low-speed machines is with a strain gauge, a device whose electrical resistance varies in proportion to the amount of strain in the device. The most widely used gauge is the bonded metallic strain gauge. It is very important that the strain gauge to be properly mounted onto the surface of the component under investigation so that the strain is accurately transferred from the test component, through the adhesive and strain gauge backing, to the foil itself. Depending on the aim of the measurements, proper location should be selected to record the proper strain. A part of the gauge itself, since these measurements are to sense extremely small changes in resistance, proper selection and use of the bridge, signal conditioning, wiring, and data acquisition components are required for reliable measurements.

The data could be also used in FEM simulations for calculations of possible outcomes for the recorded modes of operation.

These measurements are often employed to measure:

- Loads at supporting structures of power blocks.
- Operational loads at excavators bucket-wheels.
- Behavior of body of excavators during operation (stress level).
- Weighing.
- Monitoring of cracks progress at any structure.

5.3.7 Process Parameters

Process parameters[1] as condition monitoring tool got the deserved importance with the introduction of operators-driven maintenance. It is not so rear case unfortunately the maintenance staff not to be literate in reading and understanding the operations —process parameters. It requires special skills of the staff to understand the process and root causes of any changes in the process parameters. From the other side, it is very powerful tool for the proactive condition monitoring. Process parameters consist of readings in:

- Machinery outputs,
- Quality parameters (as fineness, moisture, burning effects, final dimensions of product, homogeneity of casted parts etc.),
- Pressure of fluids (gasses, liquids…),
- Temperature in gasses, liquids, or equipment components,
- Power loads of drives.

Use of the process parameters should follow the sequence:

- Customization of maintenance staff with effects of the technological process on the equipment condition.
- Customization of operators with effects of the technological process on the equipment condition.
- Determination of process parameters links to the equipment condition, e.g.
  - Differential pressure changes to blockages in pipelines and cyclones or saturation of filtering media.

[1] For purposes of this book, process parameters comprises of technological process control and quality parameters of the products.

  - Change of static/dynamic pressure to the flow or changes in the ductworking, impellers efficiency, etc.
  - Effects of the change of temperature of the process on the equipment durability (linings-refractory, liners-metal components, seals, bearings lubrication, thermal deformations, etc.).
  - Amps of running equipment links to the possible frictions, increased loads etc.
- Define simple rules for every particular equipment on monitoring process parameters from aspect of equipment condition.
- Train the staff for their use.

5.4 Proactive Information from the Condition Monitoring Techniques

The way the methods and techniques described above are mainly used is giving them property of "reactive" monitoring techniques because they are able to recognize some change once the failure mode has already "established". For example, recognition of gear teeth wear or gear cracks or wear and imbalance of an impeller is "just" recognition of already developed state. Oil degradation monitoring is again "only" showing that significant wearing of bearing or gear took place or seals were damaged and pollution was found in the oil. NDT testing also determine existence of cracks or wearing or corrosion when it has already happened and developed. Of course, the aim is not to diminish the meaning and importance of these conditions monitoring methods. In contrary, they are very important and meaningful for equipment and their condition change [36–38]. The aim is to make a difference and distinguish the **monitoring methods based on "reactive" approach** and **monitoring methods based on "proactive" approach or** how to get proactive information for monitored equipment.

For example, measuring temperature of a bearing has always certain ranges and usually no actions are undertaken until a defined limit is reached. This is more a reactive approach because when temperature reaches a limit, it is most probably a fault being generated. Following the proactive approach, the trend should be monitored and when still under the limits, but a trend is started upward, all influencing items should be checked, as cooling system, filters, temperatures indicators, etc. Here, improving cooling is beneficial by simple actions with cooling medium flow, better filtering, etc.

Another example could be vibrations measurements. Having trend toward their increase or appearance of some new harmonic in the frequency spectrum should trigger checks of its origin. With preventive/predictive inspections, nothing is still required to undertake, but with proactive approach there should be the previously mentioned check. There are cases that conditions of operation are changing as, for example, in fans to have suddenly increased dust load that will eventually lead to increased wearing of the blades and failure.

So, proactive monitoring asks for very close cooperation between maintenance and production staff and excellent knowledge of the process and nature of the faults development. Once linking all these, the monitoring becomes regular routine, or regular way of thinking.

Actually, the proactive condition monitoring is about monitoring the condition of equipment and recognize effects from the influential factors on the condition of the monitored component. It is like a step backwards toward sources of the start of condition changes in first place.

In this course, **dynamical effects** are to be used not to recognize already existing faults, but to recognize factors brining to the fault to happen. In rotational equipment, new harmonics raised due to imbalance could be sign of build-ups on the impellers or harmonics of blades that if not removed with time can lead to increased loads on the bearings and wearing of impellers. AEs can be very useful to recognize even small changes in the flow of process gasses and disturbances to be recognized that can cause later wearing or deformations of impellers or other gas flow equipment. Measurements of operational deformations of moving components can point to increased risks for starting wearing, boundary lubrications in large bearings etc.

The way the **Particles and chemical effects** are used even today in typical CM systems, if corrective actions are taken at the first signs of abnormalities to correct the root causes, they already play the role of proactive condition monitoring technique. But, if at signs of bearing wearing detected by the increased discrete particles or increased presence of particular alloy components in the oil for example, it is waited to get really damaged bearing to be replaced, than the technique is not proactive CM. If at the first signs of any degradation, investigation is undertaken and corrected (e.g., wearing of the bearing may be detected due to increased load over the same and this is improved), than the technique gets proactive role.

At **physical effects** for example, the corrosion of some components may be detected and at certain limit, component may be replaced or may be repaired by sandblasting and painting. But at proactive CM it would mean detect the causes for corrosion (fluids composition, atmospheric conditions, improper manufacturing, etc.) and correct the root causes. Another, very useful tool in proactive CM is the use of strain measurements. They can point to what is really happening in the structure of machinery and link the change of the strain to the operational conditions thus giving info on necessary (if any) changes in the operation process. For example, at overhead cranes, if strain measured on the bridges show significantly increased strain levels at operation of using the crane baskets for cleaning build-ups or lifting some overweight, these operations can be limited in order to protect the structure of earlier degradations before serious deformations or fatigue happen.

Temperature measurements are also another very useful proactive technique as it is directly linked to both: operational modes and equipment condition. Any increase can point to elevated friction processes, boundary lubrications, thermal deformations, etc. For example, increase in bearing operating temperature at same operational conditions and ambient temperature may point to change in lubrication conditions and possible phenomena of boundary lubrication that can cause final damage of the bearing. Experience is required or well established system for proactive CM to recognize the difference and react early to eliminate the boundary lubrication. Thermal stresses at rotary kilns can cause cracks in the shell in long run, and in short run, thermal crank can affect negatively supporting rollers positioning and lubrication of the roller plain bearings. By properly set system for monitoring, these effects can be avoided.

When developing condition monitoring practices (methods, techniques etc.) for certain machinery, there is a need to take into consideration that machine is a unity of elements, not just a system of separate elements (bearings, gears, shafts) and not only degradation of one specific element or fault should be monitored. It means that there is an interaction between elements and this interaction should be identified. For example, a fault in the bearing has been identified, one should answer, how this fault may influence a shaft misalignment to gearing misalignment. An influence of a faulty element on cooperation of other parts of the system should be evaluated [31, 39].

In this course, when monitoring a large bearing at conventional way used CM techniques such as SPM, enveloping, etc. would alarm at real damage, although at some at very early stages. But at proactive CM, we should investigate what factors affected the initiation of the damage as shown on Fig. 5.6 and monitor them at proper way.

In this case, we would be interested in change of load to detect increase over some limits, change in lubrication regimes (detection of boundary lubrication for example), change in exploitation terms affecting bearing surrounding or load, etc. Therefore, the techniques of proactive condition monitoring should be able to detect

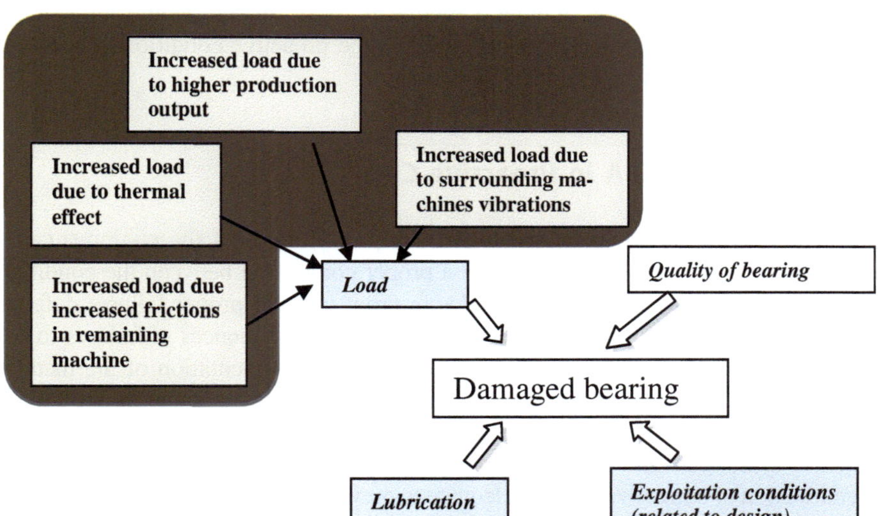

Fig. 5.6 Example factors affecting bearing damage

Table 5.5 Typical examples comparing difference between typical and proactive condition monitoring

Component	Typical condition monitoring	Proactive condition monitoring
Bearing	Vibrations (e.g., accelerometers) Temperature	SPM HD (lubrication quality monitoring) Vibration of the running machine where bearings are installed (e.g., vibrations of fan as its imbalance affects bearing health) Quality of lubrication Deflections at large shafts Exploitation conditions (cleanness, diffuse dust, elevated temperatures in the surrounding etc.) Vibrations at bearing Temperature of bearing
Gears	Vibrations (focus of GMF and sidebands) Visual and NDT at large gears	Loads at the gears Contact patterns Lubrication quality (quality of lubricant and quantity to the corresponding gears) Misalignment of shafts, housings Wear of bearings at corresponding shafts Vibrations (focus of GMF and sidebands) Visual and NDT at large gears

the effects of other influencing parameters on failure initiations and provide reliable measurements of the determined parameters (Table 5.5).

It may seem as too many activities for condition monitoring at proactive way, but it all make sense as machines of major importance and actually all listed proactive monitoring actions are multi-useful as the same have another meaning for monitoring of the condition of other components. With smart schedule and reading results and interpreting them in broader sense, the proactive condition monitoring becomes simple as any monitoring system.

5.5 Role of RCFA in Proactive CM

No matter of all condition monitoring techniques, failures of the equipment will occur and it is on the plant to establish a proper connection between the condition monitoring outcomes and failures modes in order to build proper failures defence strategy in order to minimize or even eliminate the consequences of failures to the total performance of the plant. From these reasons implementation of the method "Root-Cause Failures Analysis (RCFA)" is strongly recommendable in one Proactive and Reliability Centred Maintenance system. This methodology has actually a target to investigate and report the causes of an occurrences in order to enable the identification of corrective actions adequate to prevent recurrence and thereby protect the aspects involved (humans, health and safety, environment, equipment, etc.).

Actually, RCFA helps defining proper condition monitoring tools/techniques for tackling real root causes of failures. When conducting RCFA on particular problem or failure, the causes can be found and corrected so that the failure would not repeat again. However, in some cases, the causes cannot be simply eliminated once for lifetime as they can be combination of physical properties of the asset-technological process of the production system and humans.

The RCFA as method refers to formal failure investigation procedure initiated by failures with significant impact to the total performance of the plant. The Root Causes that are aimed to be determined are actually the cause that, if corrected, would prevent recurrence of this and similar occurrences. The root cause does not apply to this occurrence only, but has generic implications to a broad group of possible occurrences, and it is the most fundamental aspect of the cause that can logically be identified and corrected. There may be a series of causes that can be identified, one leading to another. The methodology is based on many subsequent cycles of asking "Why" something happened and going deeper with every new cycle of investigation. This series should be pursued until the fundamental, correctable cause has been identified. The investigation process is used to gain an understanding of the occurrence, its causes, and what corrective actions are necessary to prevent recurrence. The benefits this analysis brings is actually an improvement of the overall plant reliability, better knowledge to the plant personnel about all aspects of the process and reduces the risk of potential catastrophic failures.

Readers are further directed to use literature sources.

References

1. FLSmidth Institute, in *The International Maintenance Seminar*, vol I. (Copenhagen, 2005)
2. *Preventive Maintenance/Essential Care and Condition Monitoring, Preventive Maintenance/ Essential Care and Condition Monitoring*, IV edn. (IDCON, INC, 2003)
3. R.T. Buscarello, *Practical Solutions to Machinery and Maintenance Vibration Problems*, 4th edn. (Update International, Inc., Denver, 2002)
4. D. Martin, J. Van Dyke, *Integrating Vibration, Motor Current, and Wear Particle Analysis with Machine Operating State for On-line Machinery Prognostics/Diagnostics Systems (MPROS)*, © 1998 by DLI Engineering
5. Jason Mais, Scott Brady, *Introduction Guide to Vibration Monitoring: Measurements, Analysis and Terminology* (SKF Reliability Systems, USA, 2002)
6. M. Brown, *A Guide to Effective Maintenance Strategy implementation*, Asset Management Services (ABB Eutech, UK, 2010)
7. E. Becker, *Vibnode Keeps Kilns Turning* (Pruftechnik Gmbh, ICR, Germany, 2005)
8. E.B. Halim, S.L. Shah, M.J. Zuo, A.A. Shoukat Choudhury, *Fault Detection of Gearbox from Vibration Signals using Time-Frequency Domain Averaging*, IEEE (American Control Conference, 2006)
9. C.M. Harris, A.G. Piersol, *Shock and Vibration Handbook*, 5th edn. (McGraw-Hill, New York, 2002)
10. J.P. Den Hartog, *Mechanical Vibrations* (MIT, Dover publications, Inc., New York, 1984)
11. ISO17359 Condition monitoring and diagnostics of machines—General guidelines

12. ISO10816-1 Mechanical vibration—Evaluation of machine vibration by measurements on non-rotating parts—Part 1: General guidelines
13. ISO10816-3 Mechanical vibration—Evaluation of machine vibration by measurements on non-rotating parts—Part 3: Industrial machines with nominal power above 15 kW and nominal speeds 120 r/min
14. ISO10816-5 Mechanical vibration—Evaluation of machine vibration by measurements on non-rotating parts—Part 5: Machine sets in hydraulic power generating and pumping plants
15. B. Jones, *CMVA60ULS Microlog-Drill Platform Anchor Winches*, CM3050 (SKF, Gothenburg)
16. P. Johnson, S. Sarwar, *Recent Technology Trends in Machine Vibration Monitoring* (Averna Technologies Inc, Ottawa)
17. J. Mais, *Time Domain Analysis of Vibration Data: Using Time Domain Signals to Further Diagnose Machinery Health* (USA, 2004)
18. C.K. Mechefske, *Machine Condition Monitoring and Fault Diagnostics* (Queen's University, Ontario)
19. National Science Foundation, Workshop on signal processing for manufacturing and machine monitoring, University of Washington and Douglas Jones, University of Illinois, Les Atlas, 14–15 March 1996
20. B.H. Nystad, M. Rasmussen, in *Safety and Reliability for Managing Risk-Guedes*, ed by Soares, Zio Prognostics of technical condition index for an aging repairable system (Taylor & Francis Group, London, 2006). ISBN:0-415-41620-5
21. SKF Reliability System: SKF Condition Monitoring Area Center, *Vibration Diagnostic Guide*, ID CM5003 (California, 2000)
22. A. Smulders, *Time Domain Analysis*, (AB SKF, Gothenburg, 2005)
23. W. Wier, D. Trivanovic, *Using Acceleration Enveloping on Sleeve Bearings*, (SKF Reliability Systems, CM3093, USA)
24. G. White, *Cepstrum Analysis*, (DLI Engineering, USA, 1998)
25. A. Floding, Wear of spur and helical gears, ISSN1400-117, Doctoral Thesis, Royal Institute of Technology, Sweden, 2000
26. E. Rusinski, P. Moczko, D. Pietrusiak, Experimental and numerical studies of Jaw Crusher supporting structure fatigue failure. Strojniski Vestnik-J. Mech. Eng. **59**(9), 556–563 (2013)
27. Z. Stamboliska, Proactive method of low-speed machines condition monitoring, PhD thesis, Wroclaw University of Technology, Poland, 2011
28. E. Rusinski, S. Dragan, P. Moczko, D. Pietrusiak, Implementation of experimental method of determining modal characteristics of surface mining machinery in the modernization of the excavating unit. Arch. Civil Mech. Eng. **12**(4), 471–476 (2012)
29. Vibration Analysis II, Technical associates of Charlotte, PC. 1997
30. ISO13373-1 Condition monitoring and diagnostics of machines—Vibration condition monitoring
31. J.R. Ottewill, M. Orkisz, Condition monitoring of gearboxes using synchronously averaged electric motor signals. Mech. Syst. Signal Process. **38**, 482–498 (2013)
32. Wikipedia (http://en.wikipedia.org)
33. www.PDHengineer.com, *Lubrication Fundamentals*, Houston
34. Y. Kim, A.C.C. Tan, J. Mathew, B. Yang, *Condition Monitoring of Low Speed Bearings: A Comparative Study of the Ultrasound Technique Versus Vibration Measurements* (First World Congress on Engineering Asset Management, Australia, 2006)
35. E. Rusinski, P. Harnatkiewicz, G. Przybylek, P. Moczko, Analysis of the fatigue fractures in the eccentric press shaft. Mech. Syst. Mater.: Mater. Prod. Technol. **165**, 321–329 (2010). (Book Series: Solid State Phenomena)
36. E. Rusinski, Z. Stamboliska, P. Moczko, in *Proactive Condition Monitoring of Kiln Roller Plain Bearings*. X International Conference CAE2010 Wroclaw (Poltegor-Instytut, Wroclaw, Poland)

37. E. Rusinski, Z. Stamboliska, S. Milevski, in *Proactive Condition Monitoring of Low-Speed Machines in Cement Industry*. X International Conference CAE2010 Wroclaw (Poltegor-Instytut, Wroclaw, Poland)
38. G. Lind, R. Barrett, Wilcoxon Research, Inc., *Low frequency vibration measurements on a compressor gear set*, PT102 (Gaithersburg, Maryland, USA, 1993)
39. A. Jablonski, T. Barszcz, Validation of vibration measurements for heavy duty machinery diagnostics. Mech. Syst. Signal Process. **38**, 248–263 (2013)

FEM in Proactive Condition Monitoring

6

Abstract

This chapter presents application of FEM as a supporting tool for condition monitoring of existing equipments (technical assessment of condition, determination of best monitoring parameters, setting limits for early corrective actions).

Keywords

FEM · Proactive monitoring · Technical assessment · Lifetime prediction · Best monitoring parameters · Limiting values · Numerical models · Root causes · Major breakdowns

6.1 Introduction

Finite element method (FEM) is a widespread tool used commonly in all branches of science and industry. It allows conducting simulation studies, which provide wide range of detailed information on the various phenomena in the virtual space of computer systems. These simulations are designed to the best representation of the actual working conditions of the analyzed object, and thus achieve real results of virtual analysis.

In engineering practice related to the design and operation of machines and technical equipment, finite element method is used at every stage of the "life" of these objects. These stages are as follows:

(a) design stage,
(b) prototype stage, and
(c) exploitation stage.

Z. Stamboliska et al., *Proactive Condition Monitoring of Low-Speed Machines*,
DOI 10.1007/978-3-319-10494-2_6

In this chapter, the authors focus primarily on the use of finite element method in supporting and optimizing the exploitation process of machinery and equipment implemented for the operation. It is a new role of FEM, described as a tool of maintenance. It is about the use of FEM for three different cases:

- Technical assessment of existing machinery condition through equipment design and simulation of exploitation terms;
- An analysis of the components' behavior during operation in order to determine the best monitoring parameters to be used in proactive condition monitoring and setting their limiting values that should play a role of triggers for certain corrective actions;
- Determination of physical root causes for major breakdowns.

These issues will be discussed in detail in the following chapters.

6.2 FEM as a Tool for Technical Assessment of Equipment Condition

Evaluation of technical condition of machinery and equipment is an important factor in determining the accuracy and safety of their operation. Properly conducted, it provides valuable information about the current state of the object, as well as enables its forecasting. It is common engineering practice that assessment of the technical condition is often applied to a limited extent and underestimated as well as neglected. In the case of load carrying structures, only visual inspection and non-destructive tests are used. Similarly, in the case of mechanical components of machines, applied research mainly based on the perception of the human senses (hearing) or using measurements such as temperature or vibro-diagnostics are used. This approach is usually observed in the case of periodic evaluation of the technical condition. This approach, in contrast to the continuous monitoring, is the most common approach to the use of machines.

This approach to the assessment of the technical condition causes the efficiency of the process to be limited. It is related to the fact that the technical condition is defined in principle for two states such as: no defects or the occurrence of defects (crack and damage). We are losing in this way the possibility to obtain a very important information, which is the probability of damage that could be used in an exploitation strategy to provide proper maintenance.

This applies in particular to the load carrying structures of machinery and equipment, which unlike the mechanical components (bearings, gears, etc.) are much more complex, and very rarely covered by the monitoring systems. For this reason, a comprehensive evaluation of the technical condition can be an invaluable tool for enabling safe and economically feasible operation.

The authors, based on years of experience gained from the conducted research, developed a modern and unique method of evaluation of the technical condition of machinery and equipment mainly for load carrying structures of various types of machinery and equipment. This method is based on experimental studies and numerical calculations based on the finite element method.

The method allows to determine the following parameters of the test object:

- identification of areas potentially susceptible to damage,
- detailed information on the current technical condition,
- forecasting of technical condition change, mainly in terms of the phenomenon of durability depletion.

In the complex load carrying structures with a large number of structural joints, running economically justified and, above all, a quick assessment of the technical condition is practically not possible, if we apply the principle of the classical testing of all components.

The complexity of the design and loading of these objects makes it very difficult to predict highly effort areas of structure, which may be subject to accelerated degradation. In this case, it is preferable and advisable to use numerical methods such as finite element method to identify critical structural nodes.

This process involves the faithful reconstitution of the analyzed object geometry using CAD/CAE software, such as ABAQUS, ANSYS, CATIA, NX, UNIGRAPHICS, etc.

Proper modeling of the geometry is crucial here due to the fact that the presence of possible structural damage of construction is generated by local geometrical or welding related notches (hot spots).

Taking into account the specific geometric characteristics of geometrical model allow for precise determination of their effect on the prediction of present and further structure condition.

In Figs. 6.1 and 6.2, are examples of load carrying structures, structural nodes of real objects and their corresponding modeling in the 3D numerical models.

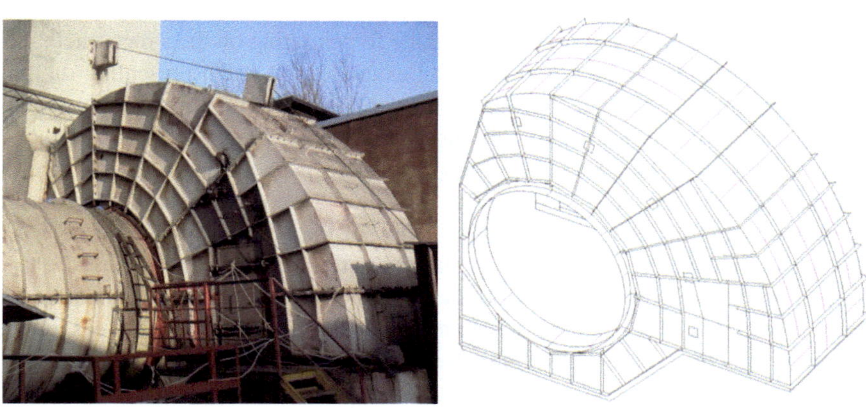

Fig. 6.1 Radial fan cover and its numerical 3D model

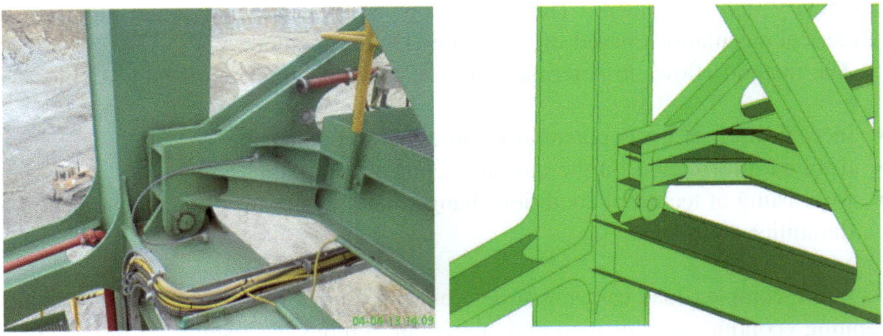

Fig. 6.2 Structural joint of load carrying structure of bucket wheel excavator KWK1500.1 and its numerical 3D model

The accurately made geometrical models are discretized with the use of finite elements and are defined in the boundary conditions in the form of restraint and loads. This phase of work is extremely important from the point of view of reliability of the results of the work, because their quality is closely related to the precision of mapping the actual working conditions of the test object.

When defining the boundary conditions, we can have recourse of the relevant standards that describe the elementary loads and associating loads combinations such as mining standards [1], hoisting [2], or multipurpose [3].

The actual conditions differ sometimes from the assumed by the authors of standards [4]. In such a case, it is advisable to carry out further identification studies on the real objects using different types of measurement methods, such as strain gages, force sensors, acceleration, displacement, and others [5–7]. With this approach, we increase the accuracy and efficiency of research.

An example of such work is identification of the real loads conditions of the bucket wheel excavator SchRs4000. For this purpose, the strain gauge monitoring system located on the bearing element of the bucket wheel drive was used. This system once calibrated, was used to record the forces acting on the superstructure of the bucket wheel excavator. In Fig. 6.3, an example of a graph presenting digging force Fs variations is shown. This graph was obtained from the monitoring system during normal operation of the excavator. Arrows show overloads of the bucket wheel drive unit, which in many cases determine the stability of the durability depletion process. The level and number of overloads must be verified with design expectations and included in the determination of the residual life of the object.

With a complete computational model consisting of a detailed geometric and discrete model as well as realistic boundary conditions (restraints and loads), we can perform numerical calculations that allow the identification of areas of the structure, which may be potentially dangerous in further exploitation (prone to failures) and should be checked for the fracture presence.

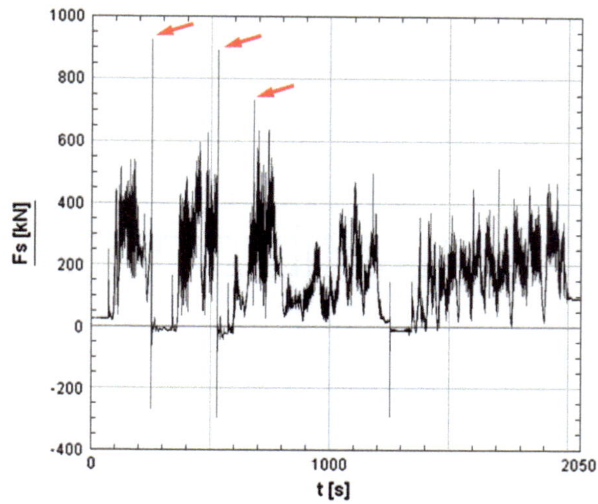

Fig. 6.3 Time signal of digging force Fs recorded during normal operation of the excavator SchRs4600—overloads of the drive unit are marked with *arrows*

Sample results of this type of analysis are shown below in Fig. 6.4, in the form of a comparison of the results of durability calculations of the supporting structure model of jaw crusher and identified faults (cracks) in the corresponding regions of the real structure.

It can be noticed that the full compliance of numerical analysis and experimental research confirms the effectiveness of the methods of assessment of technical condition developed by the authors of this work.

General scheme of numerical and experimental methods of the technical condition assessment of machinery and equipment is shown in Fig. 6.5. This method integrates virtual simulations, calculations specified by standards and experimental examinations in order to obtain detailed information about the technical condition of the object. These data are then used to predict the condition in the future. Using

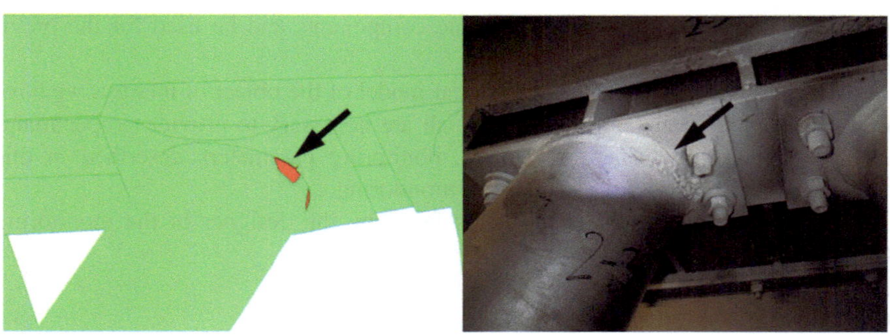

Fig. 6.4 Locations predicted in numerical model and existing cracks in the load carrying structure of the jaw crusher

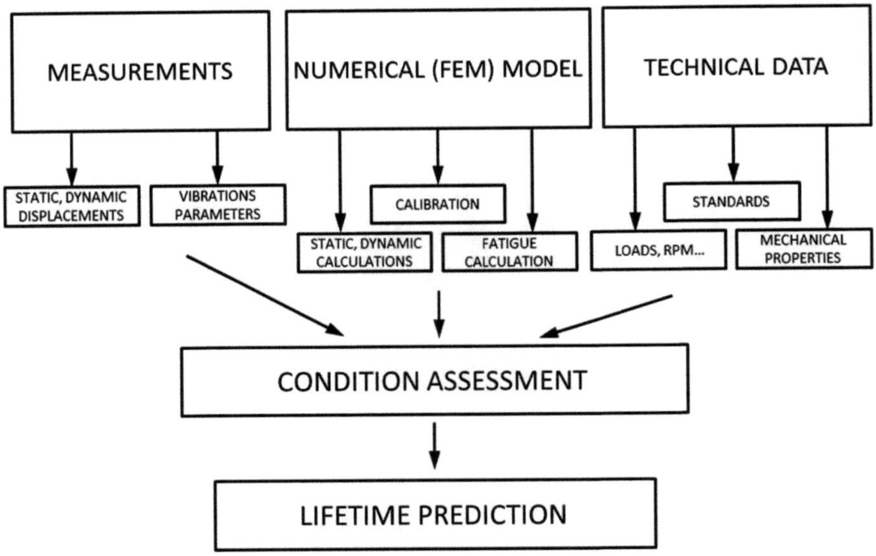

Fig. 6.5 Numerical-experimental method for assessment of technical condition

the same models and the results of the analyses, we can further determine residual life and predict how long the given object can work safely, with consideration of assumed data of operating conditions.

6.3 Determination of the Best Monitoring Parameters and Setting Their Limiting Values

In heavy duty industry, it is too "expensive" to make experiments in real conditions for determination of the best monitoring parameters, limiting values, or monitor the behavior of the equipment in some extreme cases. Therefore, numerical simulations can be employed as a tool to determine the equipment's changes related to the loads, stresses, and deviations of separate components and be used for the previously given purposes.

By creation and validation of numerical model of the object of interest, we have a tool to find out many information, which are necessary to provide proper maintenance (monitoring). There are many monitoring parameters described in this book, which can be useful in condition monitoring.

Typically, the fundamental problems faced by the designer of the monitoring system are:

- selection of the measured parameters,
- location of the sensors,
- determination of the threshold values (warnings and alarms).

Monitoring parameters are related to the working condition of the object. In the case of load carrying structures, they are mainly as follows: strain, displacement (deflection), vibration parameters (vibration velocity and acceleration).

In the case of mechanical components, there are also additional parameters related to the nature of work and the process of the facilities they serve. These are, for example: temperature, geometrical deviations, changes in magnetic properties, and other. By choosing a specific monitoring technique, we must take into account the following requirements:

- the minimum number of measurement points,
- the possibility of registration of measuring parameter (sensitivity) related to the variation of the measured signal,
- low cost per unit (e.g., per one sensor),
- possibility of transmission of measurement data to the system, taking into account the ambient conditions (e.g., temperature) and a distance to the central unit.

Taking into account the above requirements, we can build a system that delivers fast enough key information to the monitoring system, while ensuring an economically justified cost.

Most of these factors are related to the knowledge of the behavior of the monitored object in quantitative and qualitative terms. If we are able to correctly identify these conditions, we can optimally match the monitoring technique. As already mentioned, the use of numerical simulations in this case is very helpful. Having validated numerical model, we can accurately determine the optimal technique for monitoring and foremost the location of sensors.

Below is an example of the use of numerical methods for the design of monitoring system of loads and deformations of the bucket wheel drive unit of bucket wheel excavator. This unit is characterized by high power (about 2 MW) and size 12 × 8 × 1.5 m. The total mass is about 100 ton. Due to the large size and the load carried and the problems of accelerated wear of individual gear stages, it was necessary to design and introduce a system for monitoring the loads and deformations of the drive unit.

In view of the complicated construction of the transmission, it was decided to conduct numerical simulations, which allowed to determine the optimal locations and measurement methods. Numerical model of drive unit was created in such a way that faithfully represents its geometrical characteristics (housing) and internal kinematics of the individual gear stages (ratios). The numerical model of the drive unit is shown in Fig. 6.6.

The numerical model was then applied with boundary conditions in the form of loads (forces and accelerations) and restraints, representing the actual working conditions of the drive. On the basis of numerical calculations, the states of deformation, strain and stress of the gearbox were identified. Examples of results in the form of displacement and stress contours are shown in Figs. 6.7 and 6.8. These

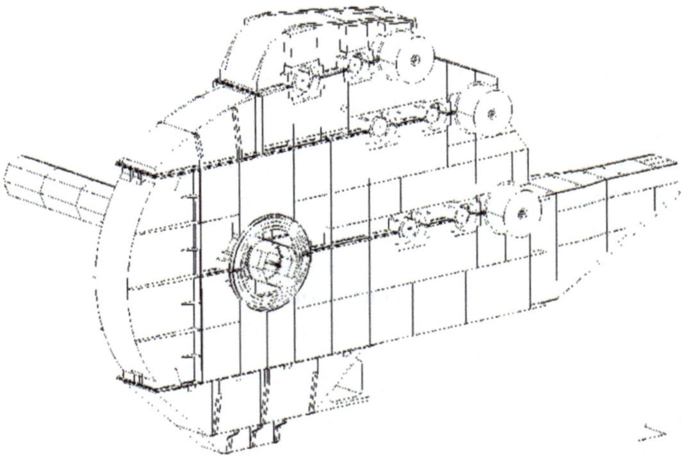

Fig. 6.6 Numerical model of the 2 MW bucket wheel drive

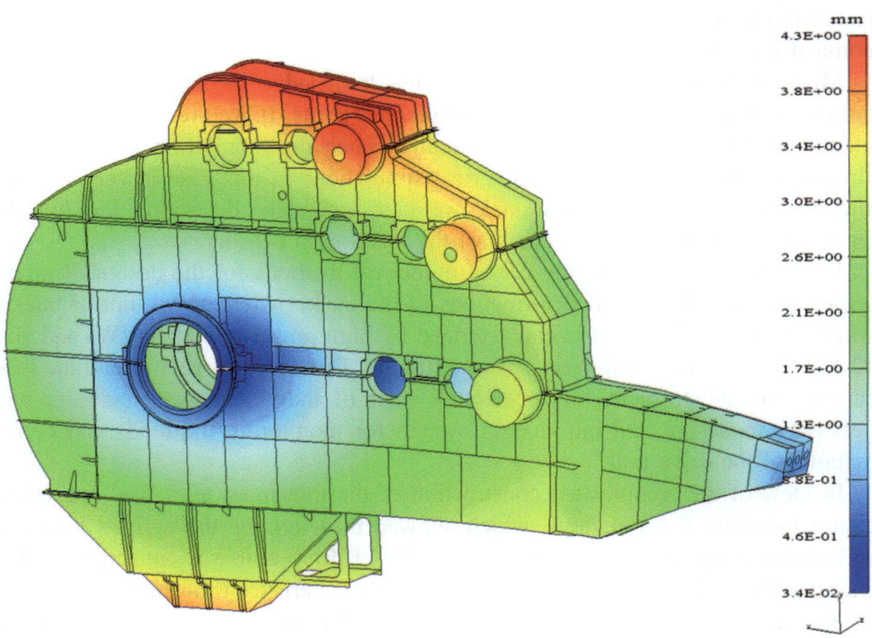

Fig. 6.7 Deformation plot of the bucket wheel drive under operational loads

results allow to accurately determine the location of sensors measuring the load (excavation force) during operation. In this case, it was decided to use strain gauges, located at the bottom of the torque beam (Fig. 6.9). This area is characterized by a high and homogeneous state effort, which provides high sensitivity of strain gauge

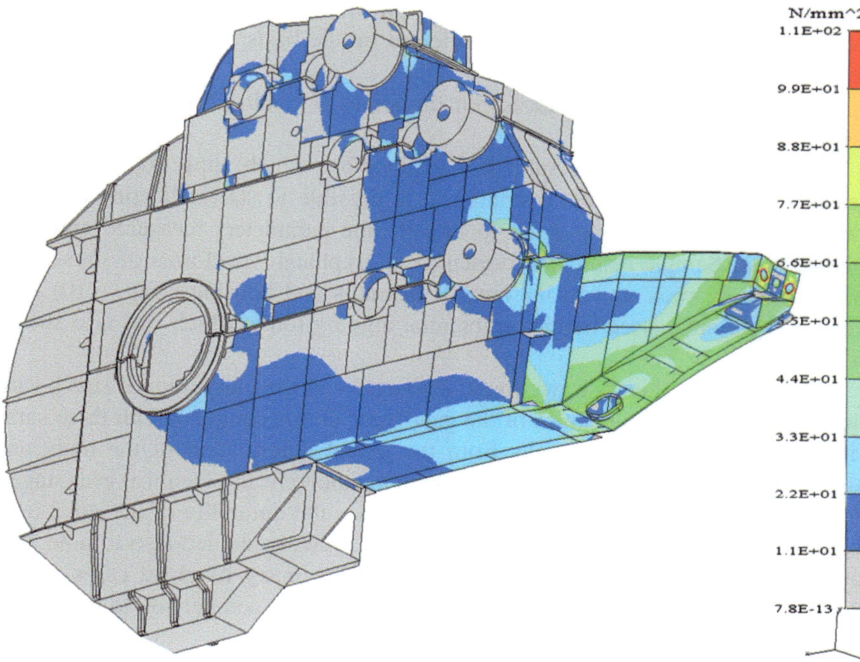

Fig. 6.8 Stress plot of the bucket wheel drive under operational loads

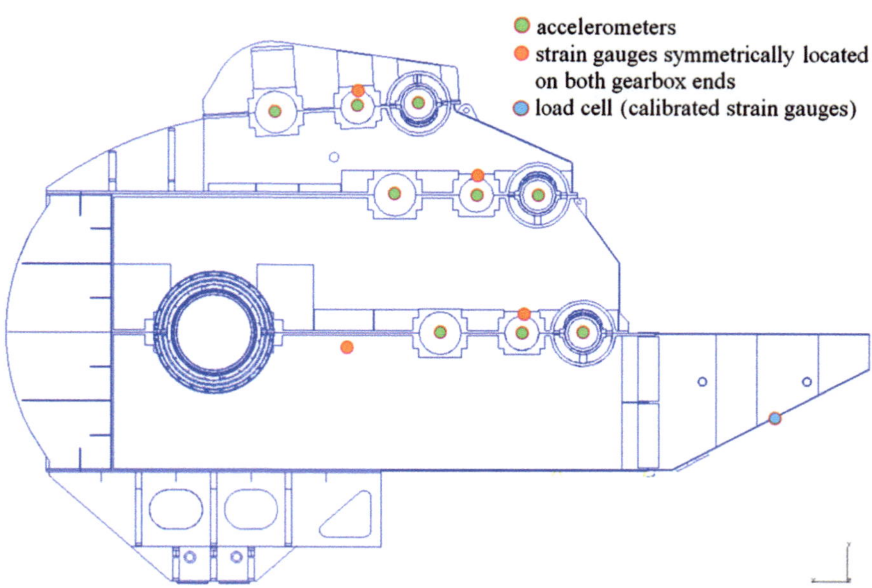

Fig. 6.9 Location of measuring points of monitoring of the bucket wheel drive

measurement. In order to monitor the deformation of the drive unit, strain and accelerations measurements at points selected on the basis of the obtained results were chosen.

The locations and types of measurement points to monitor drive unit are shown in Fig. 6.9.

The measuring system is subject to calibration, which apply to strain gauge measuring circuits. After this operation, it is possible to start collecting and analyzing measurement data in order to identify the parameters measured and their interpretation. Sample results are presented in graphical waveforms of stress variations in the gearbox unit in the area of bearings and housings (Fig. 6.10). The stress changes are caused by circumferential digging force, which was also shown on the graph.

Interestingly, in this case, we found differences in the characters of stress variations on three identical degrees of the gear (gear unit is equipped with three same drives). This phenomenon results from the global deformation of the body and causes a variation in the load transfer through the theoretically identical gear stages. This can cause local overloads, and thus influence the durability of the drive unit. The analysis of the signals from the monitoring system, enabled also to note, that during normal operation of the excavator there are many overload cases. These overloads values are higher by several tens of percent in comparison to the design values of the drive unit.

An example of this phenomenon is shown in Fig. 6.11. In the figure, there is also maximum design load indicated.

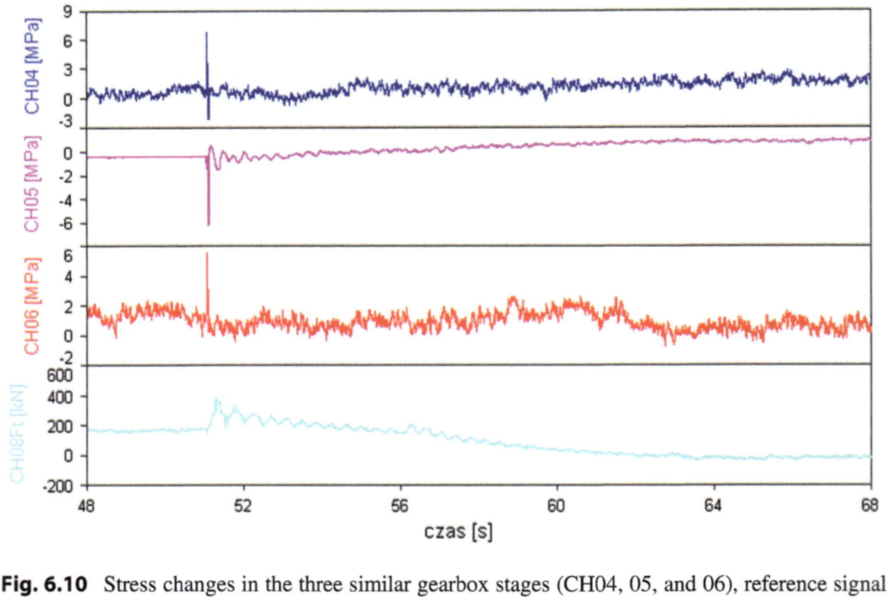

Fig. 6.10 Stress changes in the three similar gearbox stages (CH04, 05, and 06), reference signal (CH08) is shown as digging force

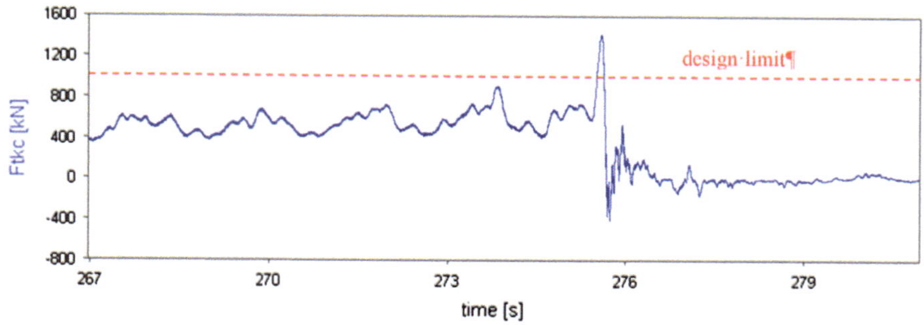

Fig. 6.11 Overload of the bucket wheel drive

Based on the results of numerical simulations and preliminary experimental tests, it is possible to determine the parameters and threshold values (warning or stop signal). It is important that detailed information on the global and local phenomena occurring in the work of the monitored object (e.g., variation in load flows through the gear and the relative deformation) can be taken into account in setting the final values (thresholds) of the monitoring system. Without having this information, which was obtained using numerical models and experimental tests, there would not be the possibility of establishing an optimal monitoring system with such a degree of safety and efficiency.

Another important advantage of the use of numerical methods in determining the threshold values for monitoring systems is also possibility to take into account changes of object, which happen during operation and their influence on monitoring parameters. These changes are mostly related to geometrical dimensions consumption caused by the process of wear and tear. Incorporating these changes in the numerical model, we can obtain information on the relationship between such consumption and the level of stress, deformation, or other applicable parameters in a given element. With consideration of obtained results, it is possible to establish proper safety thresholds.

An example of such an analysis is evaluation of the minimum thickness of the drum shell of the belt conveyor, which is shown in Fig. 6.12.

Fig. 6.12 Numerical model of the belt conveyor drum—cross section view

During the operation and subsequent repairs of the drum, the continuous reduction of its shell is observed. In order to ensure its safe operation, it is necessary to determine the value of the minimum thickness. Using the numerical model and defined boundary conditions, the spatial state of stress and deformation for a range of thickness of the drum shell were determined. Example for plot of stress and deformation distribution of the drum is shown in Fig. 6.13.

As a result of the analysis, it was possible to find the relation between the thickness of the drum shell, and the value of the stresses in the critical areas of the drum structure (Fig. 6.14). Based on this information and with consideration of mechanical and fatigue properties of steel used for drum production, the minimum thickness of the shell (g = 15 mm) was determined.

Another advantage associated with the use of numerical methods (FEM) in continuous monitoring systems, particularly in the load carrying structures is the ability to calculate and monitor the process of durability depletion.

In this approach, upon determination of the state of stresses in the analyzed element, the value of fatigue damage caused by one cycle of load or one block loads is specified. Then, accumulation of these damages is conducted, and then their durability—calculated, expressed in the numbers of cycles to fatigue crack.

A wide range of hypotheses of fatigue damage accumulation is applied. The most popular ones are as follows:

- Palmgren-Miner hypothesis [8]. The failure function is described by the following formula:

$$D_{PM} = \sum_{i=1}^{q} \frac{n_i}{N_i} = 1 \tag{1}$$

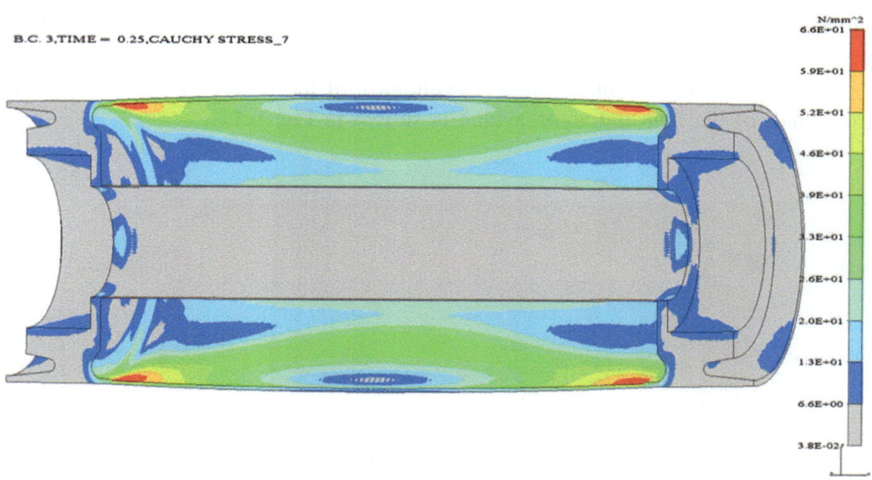

Fig. 6.13 Stress distribution in the belt conveyor drum—cross section view

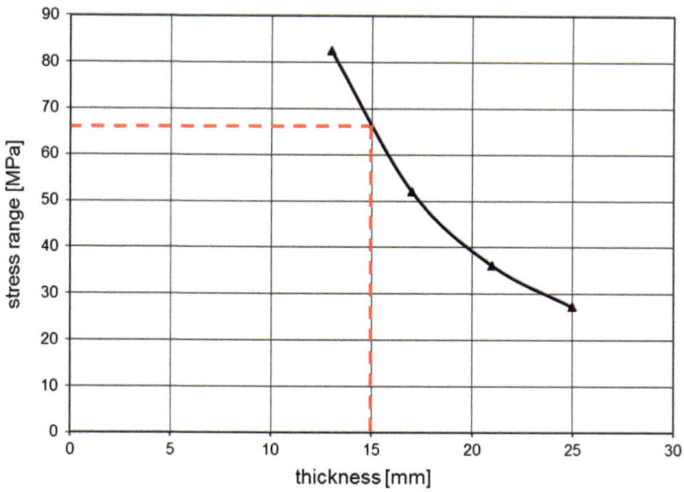

Fig. 6.14 Stress range versus thickness of drum shell, minimum drum shell thickness g = 15 mm

where:

D_{PM}	function of fatigue failure (damage) according to the Palmgren-Miner linear damage hypothesis,
n_i	number of stress cycles with their amplitude of σ_{ai},
N_i	number of cycles to failure at the stress with its amplitude of σ_{ai},
q	number of levels of stresses.

This hypothesis has a major drawback as it does not consider fatigue damages of stresses below the fatigue threshold in the accumulation.

- Modified hypothesis of Palmgren-Miner—in the hypothesis, the curve of reduced fatigue strength at the S-N graph is extended at the same angle defined as m—cotangent of the angle of the S-N graph. Modification of this hypothesis allows for consideration of loads causing stresses below the fatigue threshold.
- Haibach hypothesis [9]—extension of the curve of reduced strength fatigue at the S–N graph is made at the angle defined by the m' exponent, which is connected with the m' exponent in the following manner:

$$m' = 2m - 1$$

It is assumed that fatigue damage occurs when the sum of D reaches the value equal to 1 [10]. Cases when the value is less than 1 can also be met [11]. Cycles and semi-cycles of stress changes are normally counted using the "rainflow" method [12].

6.4 Determination of Physical Root Causes for Major Breakdowns

The use of numerical methods (FEM) in determining the causes of failures of machines and equipment is now very widely applicable.

Their use is largely dependent on the nature of the damage that occurred (sudden event or spread out over time, damage to the load carrying structure, the mechanical component, etc.). One of the more common cases of application of FEM in determining the causes of failure is estimation of stress effort and its impact on the ultimate or fatigue strength of object.

In the case, where the object on which the failure occurred is equipped with system for condition monitoring, detailed analyses of its functionality and the data recorded before and during failure are carried out. Such analyses allow to determine a set of data to run numerical simulations (for example, loads acting on object such as mechanical and thermal). With these data, and by applying them to the numerical model, we are able to analyze the phenomena occurring before and during the occurrence of failure. In this way, we can assess stress levels in a given element, the flow through the system load, deformation, temperature fields, and their gradients, and other parameters related to the event.

Examples of such analyses are shown in Fig. 6.15. Next to the object with defect (crack), there is result of numerical simulation shown in the form of stresses distributions, which are responsible for the occurrence of defects such as cracks. While maintaining the model realistic boundary conditions (restraints and loads), the accuracy of the obtained results is high. This approach allows in a precise way to determine the cause of breakdowns occurring.

Determination of physical root causes for major breakdowns has also significant meaning in the condition monitoring.

As already mentioned, the monitoring system can be very helpful in identification of the root causes by analyzing signals from the system and the use of numerical calculations (FEM).

If there is no monitoring system installed on the object, it is also possible to determine the root cause of failure in an effective way. For this purpose, the similar methodology is used, but a set of input data for computational simulations must be completed with the missing data from other sources (e.g., additional tests, the data specified by standards concerning loads, restraints, and other assumptions).

After conducting such analyses, it is relatively easy to introduce such a system, which will monitor condition of an object prone to failure areas "discovered" by the fracture and confirmed for reasons in the numerical analyses. Obviously, setting safety limits in such a case is simple task.

Corrective action to restore the object to the condition before the failure, may therefore be the first step to be followed by the corrective actions, that will prevent accidents in the future.

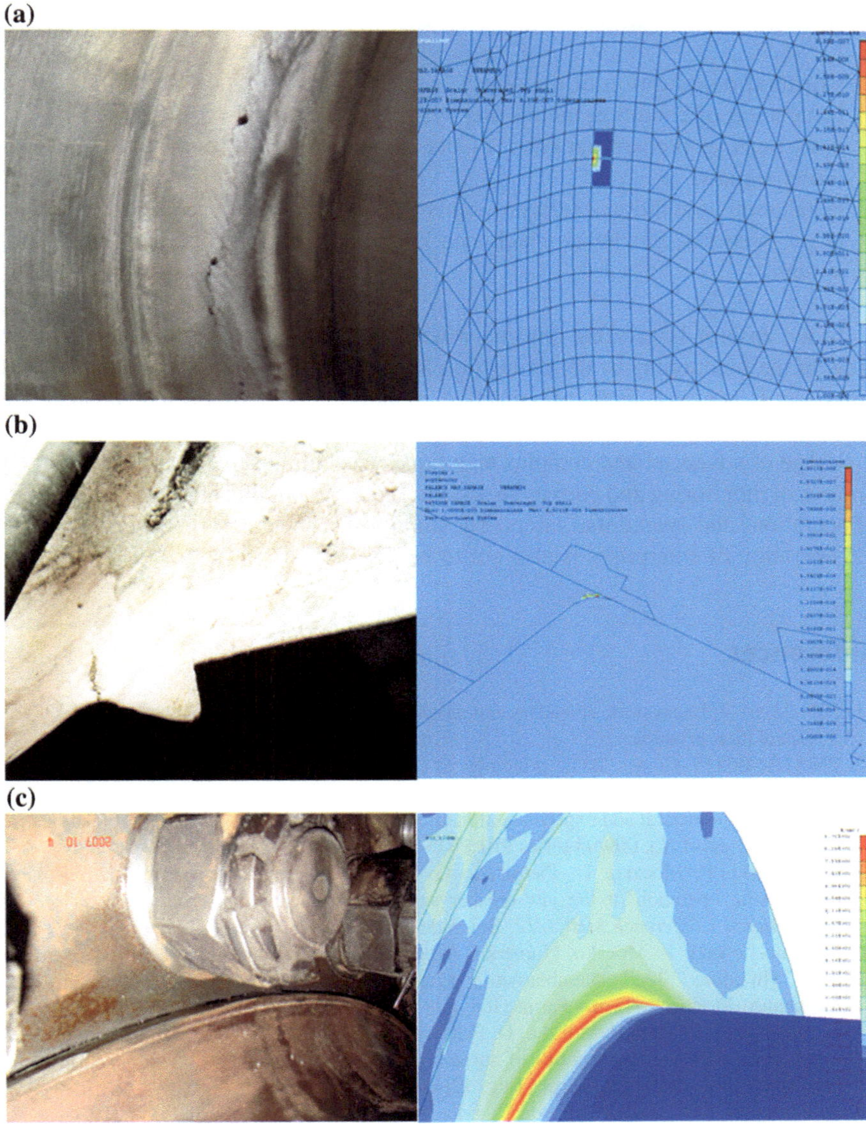

Fig. 6.15 Comparison of the accuracy of fault location prediction in a numerical model and real object with defect: **a** eccentric press shaft, **b** structural joint of the bucket wheel boom, **c** hollow sleeve coupling

Among these activities, the following can be mentioned:

- structural modification of the object (e.g., redesign of critical structural nodes),
- installation of condition monitoring systems,
- limitations of process parameters—determination of safe thresholds for monitoring signals,
- installation of additional safety systems against the occurrence of further failures (e.g., overload clutches), which can be controlled by condition monitoring system.

In the case of large-scale objects, where operation is critical from the point of view of the functioning of the production lines or entire companies, comprehensive approach is required both to proactive monitoring and to resolve breakdowns occurred.

Usage of advanced and complex techniques (including FEM), which support the process of monitoring and corrective actions, increases significantly the effectiveness of operations and operational safety. The impact of such actions on the economics of the operation of enterprises is also not to be underestimated.

References

1. DIN 22261-2, Excavators, spreaders and auxiliary equipment in opencast lignite mines—Part 2: Calculation principles
2. PN-87/M-06515, Cranes. Basic principles of dimensioning of steel lifting structures
3. Eurocode 3. Design of steel structures Part 1–1, ECS (1992)
4. P. Moczko, Numerical and experimental method of prediction of the durability of load-bearing structures, Report No. 009/03 series PREPRINTY, Wrocław University of Technology (2003)
5. E. Rusiński, P. Moczko, D. Pietrusiak, G. Przybylek, Fatigue resistance and resonance problems in operation of the jaw crusher supporting structure, in *Proceedings of the XX International Conference MHCL'12*, Belgrade, 2012
6. J. Czmochowski, P. Moczko, Numerical-experimental analysis of bucket wheel ex-cavator body vibrations. 22nd Danubia-Adria symposium of experimental methods in solid mechanics, Monticelli Terme-Parma, Italy, 28 Sept–1 Oct 2005, pp. 294–295
7. E. Rusinski, S. Dragan, P. Moczko, D. Pietrusiak, Implementation of experimental method of determining modal characteristics of surface mining machinery in the modernization of the excavating unit. Arch. Civ. Mech. Eng. **12**(4), 471–476 (2012)
8. A. Palmgren, Die Lebensdauer von Kugellagern, VDI Zeitschrift **68**, 339–341 (1924)
9. E. Haibach, Modifizierte lineare Schadenakkumulationshypothese zur Berücksichtigung des Dauerfestigkeitsabfalls mit fortschreitender Schädigung., Technische Mitteilungen no. 50/70 der Lab. Für Betriebsfestigkeit, Darmstadt Germany, 1970
10. H.T. Corten, T.J. Dolan, in *Cumulative Fatigue Damage*, Proceeding of the International Conference on Fatigue Damage, Institute of Machanical Engineers (London 1956), pp. 235–245
11. C.M. Sonsino, Overview of the State of the Art, in *On Multiaxial Fatigue of Welds*, V International Conference on Biaxial/Multiaxial Fatigue and Fracture (Kraków, 1995)
12. Y. Murakami (ed.), *The Rainflow Method in Fatigue* (Butterworth-Heinemann, Oxford, 1992)

7 Case Studies of Proactive Condition Monitoring Applications

Abstract

This chapter presents practical case studies of applying the new proactive approach for condition monitoring of a cement rotary kiln as a typical example of low-speed machines. All required steps are given for the preparatory phase and the ongoing cycle of activities together with recommended proactive and corrective measures. Few real cases recorded in industry are given pointing to the proactive side of such condition monitoring system.

Keywords

Proactive condition monitoring · Cement rotary kiln · Supporting roller · Measurement method · Measuring system · Eddy current probes · FEM · Faults · Hot bearing · Roller out-of-roundness · Kiln crank

7.1 Application for Cement Rotary Kiln

Machines similar to cement rotary kiln are often employed in the heavy duty industry and the cement industry is a representative example of this industrial branch. Following the principles of proactive approach for condition monitoring (as shown in Figs. 4.3 and 4.4), the rotary kiln was analyzed and described for its functions, equipment components, and FMECA analysis was made as given in the example in Chap. 3. Further, the kiln was analyzed using the procedure of flowchart for the new proactive condition monitoring method and the same is given again in Fig. 7.1 with small additional comments on the peculiarities related to the cement rotary kiln.

The kiln was analyzed for its primary and secondary functions. The rotary kiln has a major function of producing clinker, the core component of cement. But at the same time, there are secondary functions to provide the whole process to be safe for people and equipment, to provide lowest emissions in exhaust gases, to enable optimization of the burning process, etc.

Z. Stamboliska et al., *Proactive Condition Monitoring of Low-Speed Machines*,
DOI 10.1007/978-3-319-10494-2_7

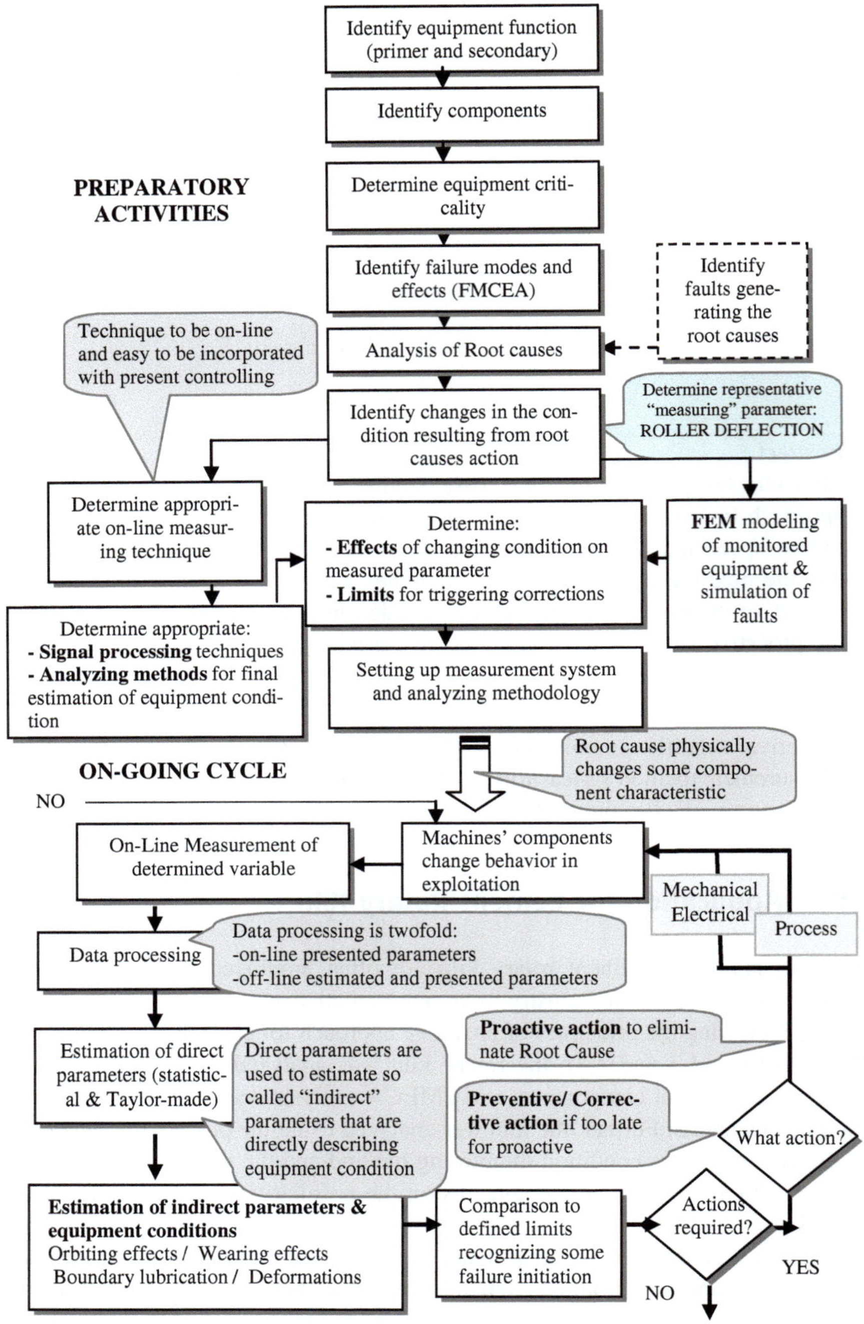

Fig. 7.1 Principle of new condition monitoring method

Kiln components were mainly identified as the following:

- Kiln inlet with sealing system.
- Kiln cylinder.
- Kiln drive (main and auxiliary drive with open gear transmission in most cases).
- Kiln outlet with sealing system.
- Kiln hood.
- Kiln supporting stations (tyres and rollers).

The criticality of the components analyzed are not discussed here, but we just point out that the highest criticality are the supporting stations, open gear transmission with the auxiliary drive, and kiln cylinder.

Failure modes and their effects as well as their criticality have been determined (see Table 2.2 in Chap. 2), as well as the root causes.

Further to this, in Chap. 2, examples were given for two major failure modes of the rotary kiln and how to determine the most appropriate condition monitoring technique. It was estimated that if the appropriate representative monitoring parameter is found, proactive monitoring would be the best. The rest of the chapter follows the process of determining the monitored parameter, establishing measuring system and tailor-made tools for estimation of the kiln condition beyond the mentioned two major failure modes.

Following the principle of determining maintenance strategy based on failure modes, it has been pointed out that for all major modes, there are reactive condition monitoring methods and techniques and for some even proactive. The highest risk number was given to two of the major failure modes, the kiln crank and boundary lubrication of the roller plain bearings. The reason was severity of the consequences, but also the inability to detect them early enough with existing CM methods. Further, the kiln was analyzed and put through the steps of the proactive condition monitoring approach as described in Chap. 5. The target was to find a way to monitor some parameters of the kiln that would act as proactive monitoring factor for the kiln crank and bearing boundary lubrication. Going from the beginning to the end of the preparatory phase, the highest attention was paid to determination of a representative monitoring parameter for the proactive approach and how to get reliable indications for the monitored condition change. The work resulted in pulling out one parameter that had the following benefits for the proactive approach (refer to Sect. 4.2) [1]:

- Reliable link exists between the effect of the failure mode on the equipment component condition change and the monitoring parameter.

Kiln crank as failure mode affects the kiln tyre rotation causing “orbiting” of the kiln section and tyre, whose range depends on the crank magnitude. The kiln becomes eccentric [2, 3] changing the load to the rollers, or there is a cycle of load being transferred to the rollers (Fig. 7.2) resulting in change of rollers deflection within the fixed housing.

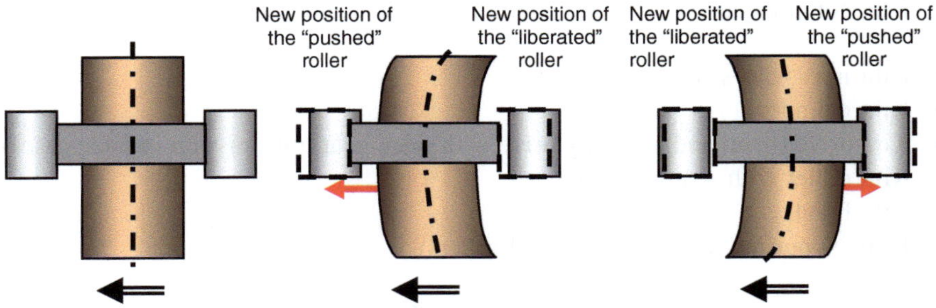

Fig. 7.2 Kiln crank effect on supporting rollers deflection from normal position

- Easy recognition of change in monitored parameter value.

When load is changing from one to the other roller, the rollers are pushed outwards when the convex section come in "touch" to the corresponding roller, and roller is liberated from the main load when the convex section is at opposite side of the roller, thus roller is moved back under the kiln (see Fig. 7.2).

- Feasibility of the monitored parameter change measurement.

Since we have determined a movement that is evident on the external side of the whole kiln tyre rollers assembly, the question of measuring was actually to find answers to:

- What to measure: displacement, vibration, or acceleration?
- Would there be suitable measurement techniques to sustain the heavy duty conditions in the kiln, as installing any probe inside a roller housing in existing set is not a good solution.

Cement kiln is a low-speed rotary equipment. The kiln rotates at about 2–7 rpm and rollers usually rotate at speed in the range of 2–5 times of the kiln (it is determined with ratio between the tyre and roller carrying surface diameters). Based on the many sources [2–4], examination [1, 2, 3, 5], and discussions on low-speed machines in Chaps. 3 and 4, displacement measured at supporting roller should be the best describing parameter for low-speed rotary kiln.

Using displacement, the rollers movement could be described by absolute values ($S_{\text{peak-to-peak}}$, S_{MIN}, and S_{MAX}), and relative change of displacement values in time. Converted into the physical process on-site, displacement measured is actually the roller deflection in radial direction of the tyre.

The measurement technique employed would have to sustain the harsh environment of the kiln, such as the polluted ambient with abrasive dust particles, oil leakages, heat radiations from the kiln shell, dynamical effects from the surrounding machines (vibration impacts from the cooling fans, gearmeshing at the open gear transmission, drive gearbox, etc.). In addition, the measuring probes would be best to be installed outside the roller assembly in order not to influence the equipment integrity. Due to rotations and relatively high deviations of the rollers surface, tyres run out that in some cases can be up to a few millimeters and difference in the dimension in cold and hot (difference of about 2–3 mm), employment of contact probes is not a good solution. Last, but not the least, the frequency response of the probe should suit the speed and amplitude of roller deflections. The answer was found to noncontact eddy current techniques and probes with corresponding performances that would measure deflections via roller carrying surface.

- Establishment of a measuring system without large costs and its effective implementation in the existing controlling system.

Systems of eddy current probes, drives, cabling, and overall technique are available on the market and there are a variety of products. Installation of a full set for one rotary kiln with three supporting stations would cost about 15–20,000 €, which is less than 8 h lost production of a small size kiln (1,200–1,500 tpd). Drivers can give simple signals 4–20 mA and then the rest can be set via PLC in the controlling system and processed further as we want.

- Obtaining final parameters that are easy to understand by the operator and maintenance staff.

Measured values of rollers deflection by themselves are easy to be understood, and by setting limits it would be easy for operators to react by simple comparison of values. In addition, data as absolute values in certain preset sampling time can be processed and analyzed with some statistical processing methods or FFT. Again, final results are not difficult to be understood by maintenance engineering staff.

> Summarizing all the above, it was concluded that the rollers deflection, expressed as displacement from a fixed point, is a confident and reliable parameter to serve as a base for proactive condition monitoring of the kiln.

The principle of the adopted method is given in Fig. 7.1. It follows the basic diagram in Chap. 5. The theory of recognizing conditions for development of any potential failure is as follows: when a change starts developing in the equipment condition, the kiln components start behaving differently. Accordingly, their describing parameters, such as temperature profile, position, run out, contact pattern, etc., start changing. This of course affects in some way the selected monitoring parameter, the roller deflection. Deflection can change in its MIN, MAX, or total value depending on the tyre run out, crank degree, temperature distribution, and some other parameters [2, 3].

Based on such change with predefined algorithms (by comparing direct parameters with their physical meaning), real kiln states can be estimated, such as kiln crank, tyre radial run out, conditions for boundary lubrication, out-of-roundness, orbiting, etc. Following this, the actual system for monitoring was designed and installed in a cement rotary kiln in a factory [6, 7].

The positive experience is given in this book including separate case studies of faults indication recognition. In this case, the system was measuring deflections of the rollers at the 2nd and 3rd stations of a 3-stage kiln. Signals were transferred to the control room in separate databases for analysis. Details on the measurement system and analysis algorithms are given as follows.

7.2 Measurement Methods and System

In order to have a complete set of all valuable information for the kiln condition, a new measurement system was installed for rollers deflection measurements, while some data were taken from existing measurement systems such as for kiln shell temperature and bearing temperature. The latter two systems are considered for this purpose only as auxiliary systems and they will be briefed in the following.

In order to create a suitable system for measuring roller radial displacement for required measuring range, accuracy, and also appropriate for the machinery working environment, the following were the criteria:

1. Capability to measure radial displacement of rotating equipment of size of 1–2 m in diameter made of steel with low rotating speed of less than 10 rpm.
2. Measurement method to be noncontact due to complexity of the machinery.
3. Measuring equipment to have wide measuring range to comprise the thermal changes of the kiln at cold and hot conditions (estimated range 4–5 mm).
4. Measuring equipment to have high accuracy to detect any changes that could be reflected in the plain bearing gap reduction (0.01 mm).
5. Measuring equipment to be resistant to harsh environment conditions (temperature variations of 50 °C, dust, oil drops, and vibrations as main influencing factors).
6. Measured results have to be capable to be transferred to a distance of about 100 m and be incorporated in existing monitoring equipment in the kiln control room (CR).
7. Have possibility for visual interpretation on the measuring results in real-time.
8. Have possibility to make further analysis of measured data.

Based on these requirements, a configuration was designed as presented in Fig. 7.3.

Eddy current probes were employed for deflection measurements [1] (with sensitivity of 3.6 mV/μm nominal and linear range 0–5 mm/1-linearity of 1 %).

In order to obtain the best representative data, measuring probes had to be installed under the same angle of the load transfer to the roller from the tyre (30°), Fig. 7.4.

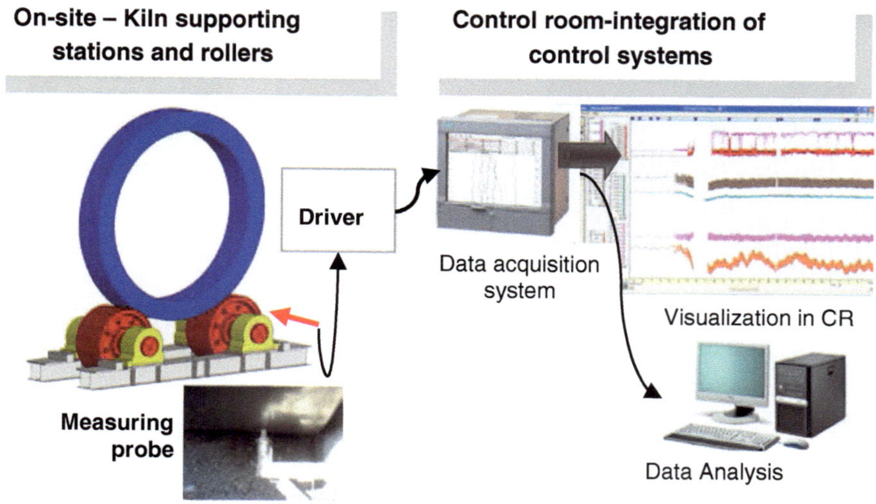

Fig. 7.3 Configuration of the measuring system measuring system

Fig. 7.4 Installation of measuring probe

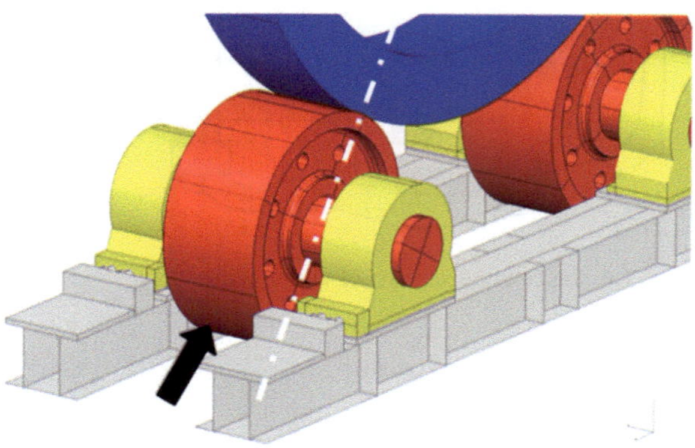

The probes were installed with special robust holders and whole measuring set was provided with IP55 protection. Cables and probe were additionally protected with cover against human influence and coarse particles. The scheme of installation is presented in Figs. 7.4 and 7.5. In total, four probes were installed for monitoring four supporting rollers at two stations (station nos.2 and 3 as stations with the major faults) as presented in Fig. 7.6.

7.2.1 Data Acquisition System

Signal from the drivers (4–20 mA) was transferred to the control room (CR) and collected by display recorder [8] (Fig. 7.7) and also stored in a secure binary

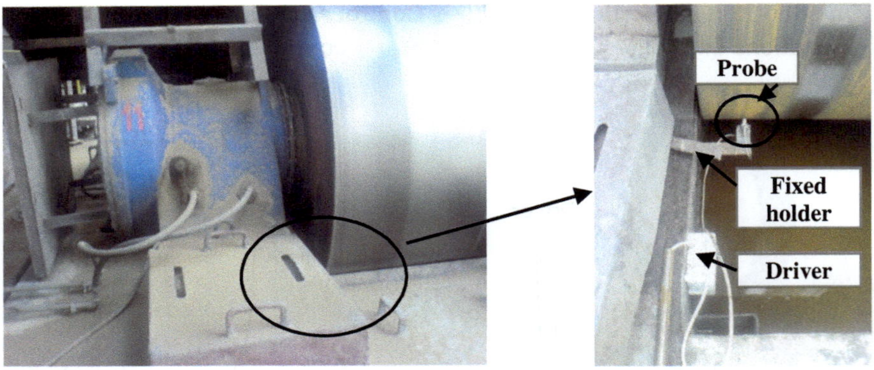

Fig. 7.5 Installation of measuring probe on-site

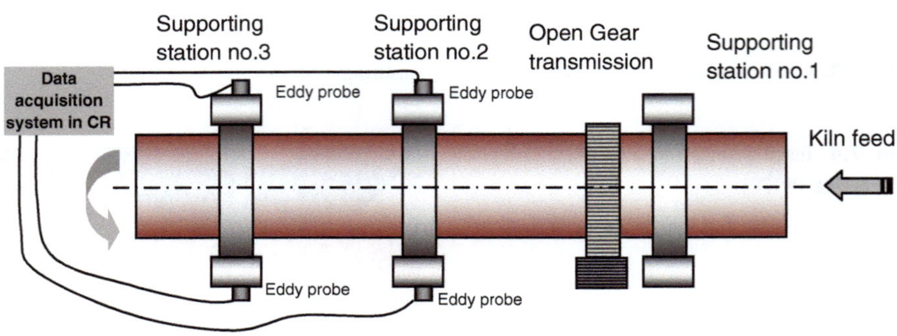

Fig. 7.6 Installation scheme of eddy probes

Fig. 7.7 Data storage system

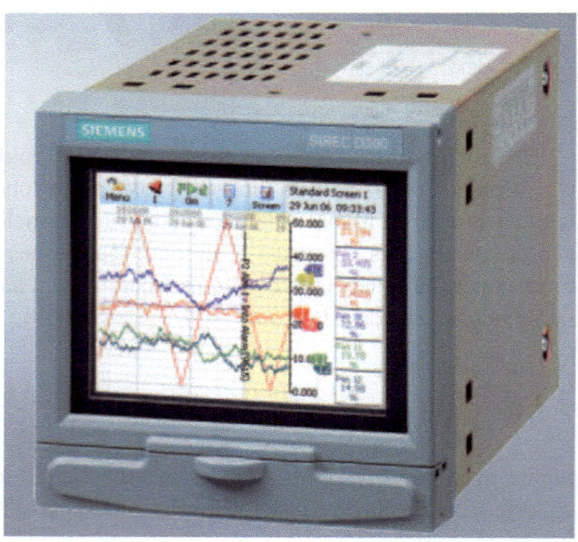

encrypted format, with the recorder's configurations, providing added security of the data files. Data transferred from the measuring probe were gathered in database *.dat and they were convenient for further analysis by different software. The system is incorporated in CR where it was accessible to operators.

7.2.2 Auxiliary Measurement Systems

In addition to the main measuring system, a kiln shell scanner (functioning principle shown in Fig. 7.8) and temperature measurements of the roller bearings by platinum resistance thermometers PT100 probes (Fig. 7.9) were used.

The signal from the probe1 was transferred to PLCs in the control rooms. Values could be read on a screen and also a history trend line could be graphed over the last 4 weeks for all roller bearings. Trends used for this example [1] are shown in Fig. 7.10.

7.3 Data Analysis Techniques

In order to enable estimation of the kiln condition, measured data had to be analyzed further. Considering the final parameters we would like to obtain, linked to the real physical properties of the equipment behavior on-site, the following techniques were anticipated to be most useful for data analysis:

- Visual presentation of data (time trends, actual values).
- Statistical data processing: estimation of values S_{MIN}, S_{MAX}, $S_{peak\text{-}to\text{-}peak}$, Least Square method for fitting curves.
- FFT of the signal and analysis of corresponding harmonics.

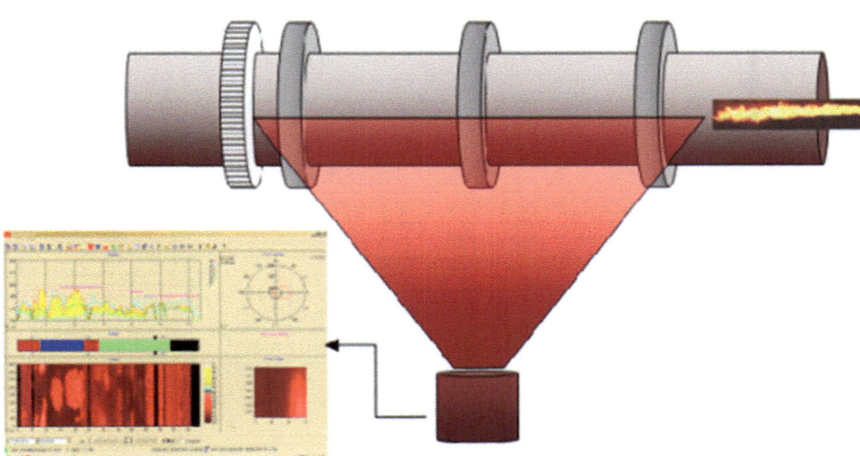

Fig. 7.8 Kiln shell scanner [3]

Fig. 7.9 Temperature measuring installation of roller bearing by PT100 probe

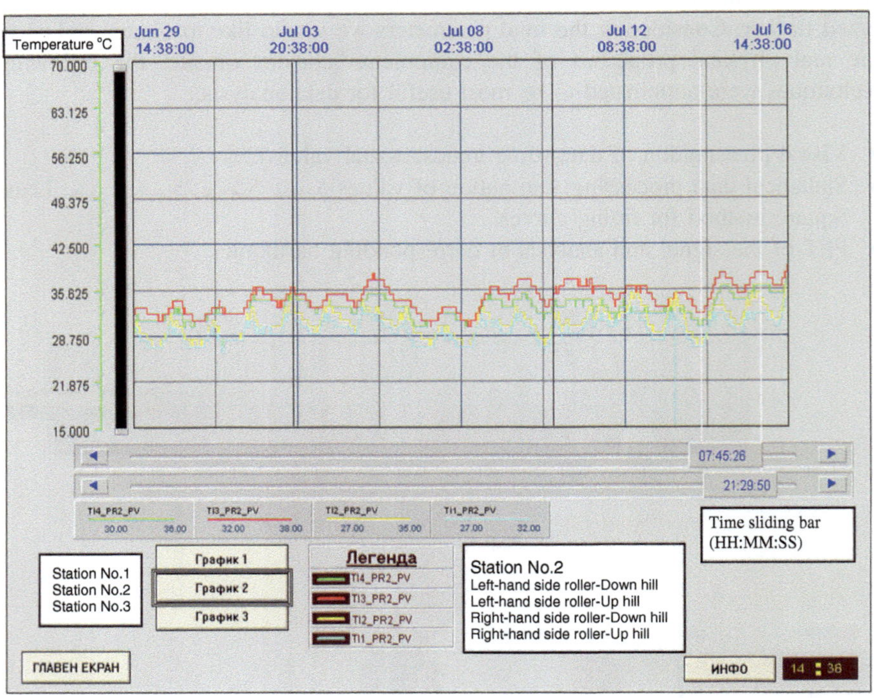

Fig. 7.10 Visualization of roller bearing temperatures

7.3.1 Visual Presentation of Data

Visual presentation technique is actually the first online tool to be used by operators. The technique represents visualization of roller deflection value and their change in time. Data are presented together with kiln amps and kiln speed (number of rotation) and they can help the operator decide on actions. General visualization used in this case study is presented in Fig. 7.11.

On the ordinate, the values of the following parameters are shown:

- displacement in micrometers,
- kiln load in Amps,
- kiln revolution in rotations per minute.

On the *abscissa*, time expressed in months/dates/time is shown with scaling possibility. The change in time domain shows the trend of the deflection as well as the general change under different conditions (kiln stoppages, effects of daily temperature, etc.).

Apart from the trend, this software was enabling ZOOM technique, where all single sampling points could be seen and exact values of deflection read. Such ZOOM technique also provided information about deflection change over one revolution of the roller, but also over one kiln revolution. This is valuable information because a conclusion can be made if there is any change in roller behavior coming from the rollers itself or impact coming from the kiln. Figures 7.12 and 7.13 show two such examples.

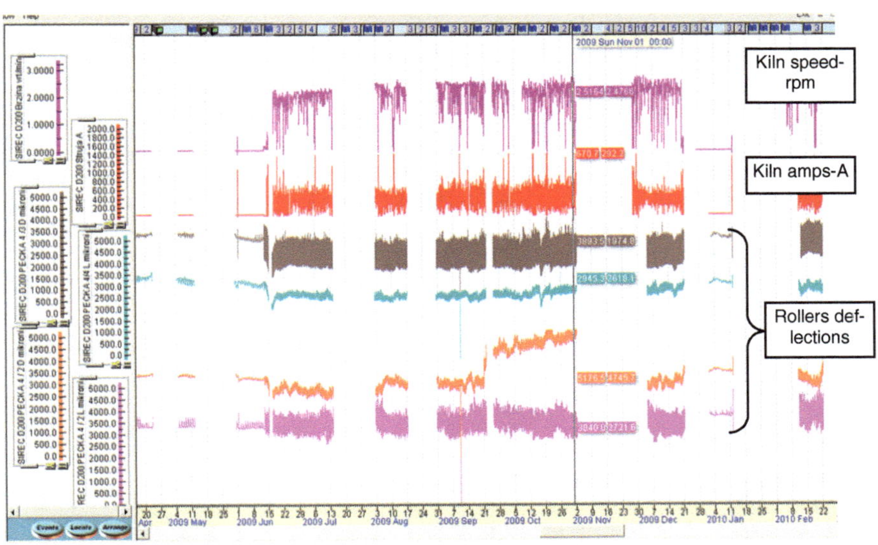

Fig. 7.11 Values and change of rollers deflection in time

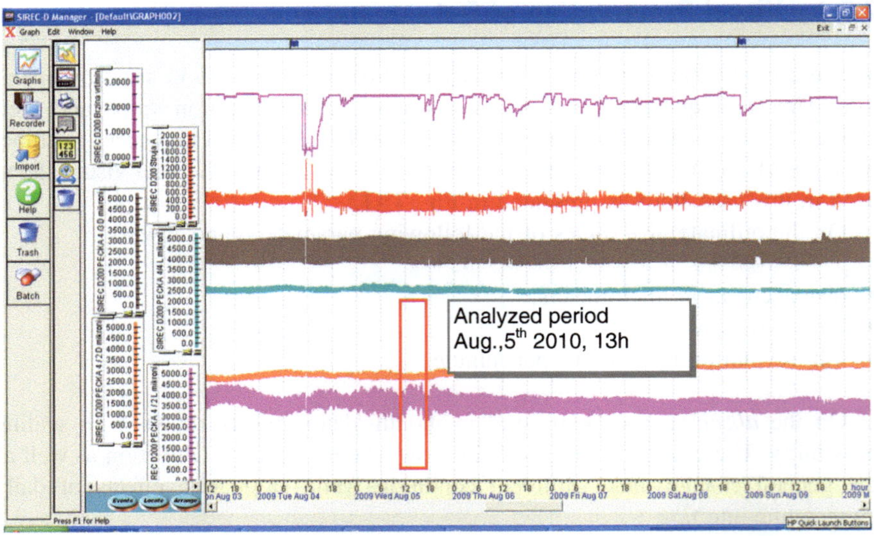

Fig. 7.12 ZOOM function in analyzing causes of roller deflection changes

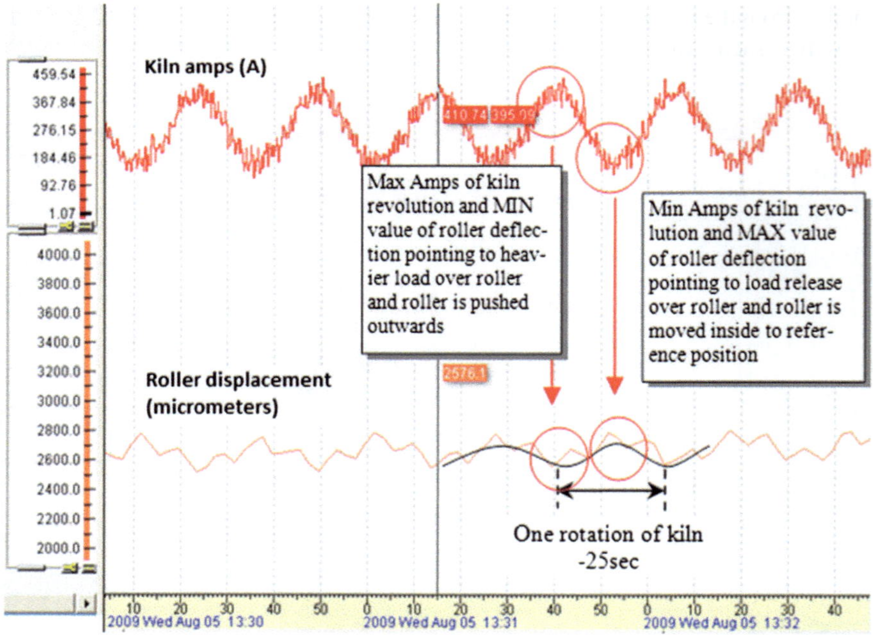

Fig. 7.13 Detection of significant kiln influence on the roller deflection change with ZOOM function

7.3.2 Statistical Data Processing

This tool is provided by exporting data into *.dat file and analyzing them by any software working with such data to estimate parameters as MIN value of deflection, MAX value, $S_{\text{peak-to-peak}}$, $S_{\text{mean values}}$, etc. Statistical estimations are used for detecting following parameters at the monitored equipment:

- **Roller max loaded position** $S_{MIN} = \text{MIN}(S_i, S_{i+n})$; n is number of sampling points estimated over several kiln revolution
- **Roller liberated position** (smallest load) $S_{MAX} = \text{MAX}(S_i, S_{i+n})$; n is number of sampling points estimated over several kiln revolution
- **Cyclic behavior of the kiln (radial run out)** $S_{\text{peak-to-peak}} = S_{MAX} - S_{MIN}$, estimated over several kiln revolution (without filtering "out-of-roundness")
- **Roller out-of-roundness** $S_{\text{out-of-roundness}} = S_{MAX'} - S_{MIN'}$, estimated over one roller revolution (noted by ') at empty kiln and lowest possible speed and by *method of least squares* for fitting curves
- **Real kiln crank** by fitting SINE function in the measured data over several kiln revolutions (filtering "out-of-roundness")

Physical meaning of these parameters is shown in Fig. 7.14:

The method of Least Squares was mentioned twice and it has proved a very useful tool for adjusting the parameters of a model function (in this case, a circle and a SINE function) to best fit a data set when estimating out-of-roundness and kiln crank (SINE function).

7.3.2.1 Fitting SINE Function

Fitting SINE function is used for analyzing the kiln effect and to determine the "pure" influence of the kiln crank over rollers deflection by removing the influence of rollers' "out-of-roundness." It is known that the kiln rotation is periodical

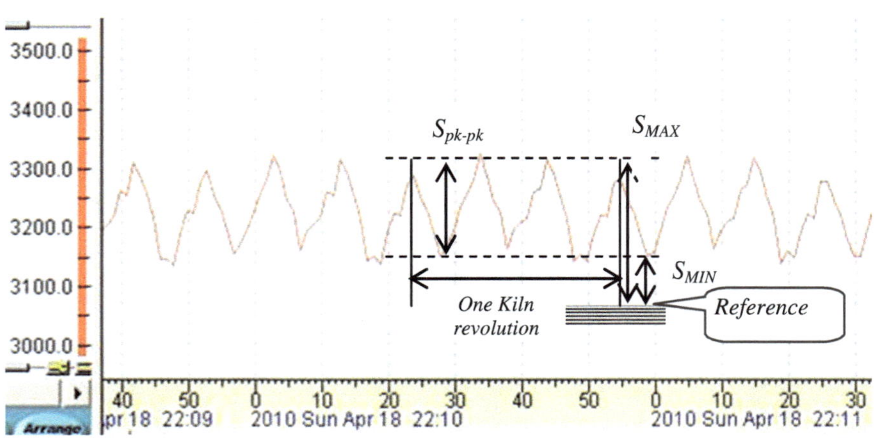

Fig. 7.14 Physical meaning of statistical parameters

movement with frequency (ω) equal to its rotations and can be represented by Eq. (7.1). So, in this case having the measurements performed, using deflection measured, the number of rotations of the kiln in the amplitude A and phase α is all that needs to be determined.

$$y_i = f(x_i, \beta) = A \sin(\omega t_i + \alpha) \tag{7.1}$$

where:

- t_i is the time sequence of sampling (1 s in our case)
- y_i is the roller deflection measured in μm.

7.3.3 FFT Analysis of the Signal and Analysis of Corresponding Harmonics

FFT analysis was used for offline analysis for confirming the influences of the kiln (tyre) or roller and detecting the root cause of the increased overall deflections. The analysis enables to distinguish the main harmonics of the kiln and rollers. The signal from rollers deflection can be analyzed as a set of periodic signals with their own frequencies and magnitudes. The FFT graph shows separate harmonics originating from the kiln (frequency corresponds to kiln revolution) and rollers (frequency corresponding to rollers revolution). A baseline measurement set is made as a reference representing kiln normal condition with even and regular loads and without any operational disturbances in period after kiln overhaul when bricks are still new and not worn out. Such reference FFT diagrams have been determined for all 4 rollers. Example for station no.2 is given in Figs. 7.15 and 7.16.

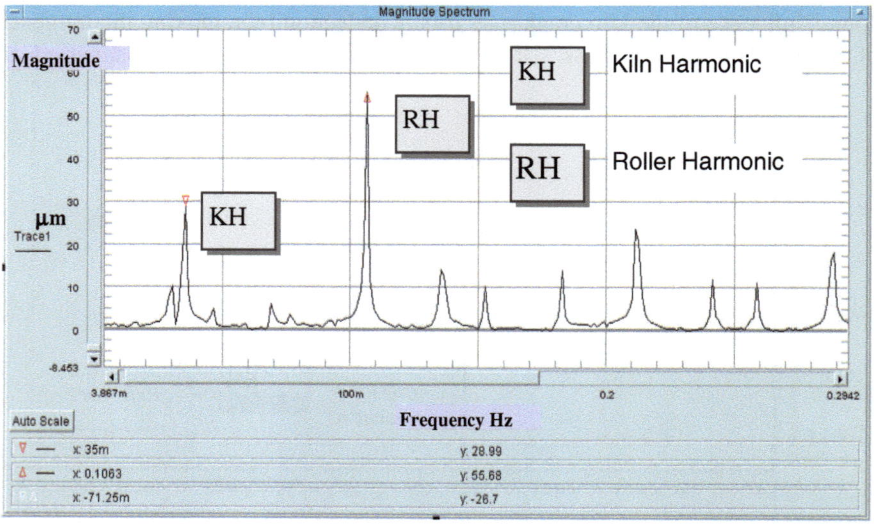

Fig. 7.15 FFT analysis of 2nd station, *Left-hand-side* roller during normal kiln operation

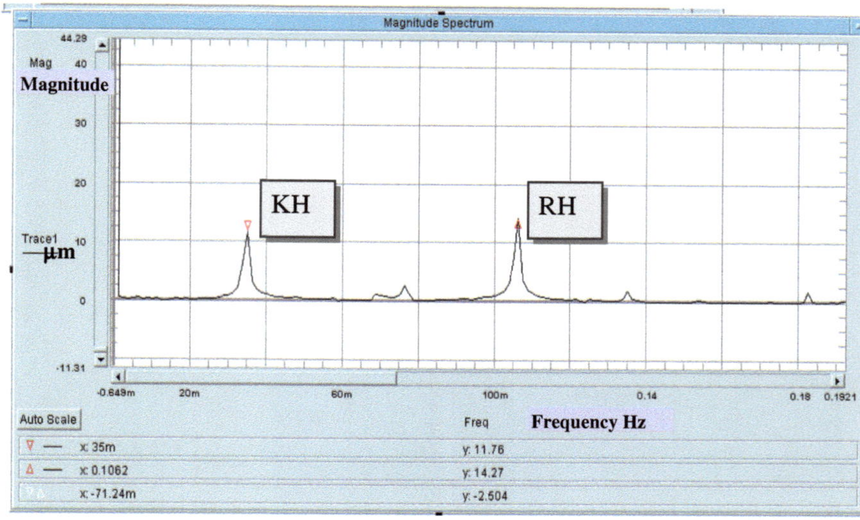

Fig. 7.16 FFT analysis of 2nd station, *Right-hand-side* roller during normal kiln operation

Harmonic representing kiln rotation is in the range of 30–40 mHz (rotations 2–2.5 rpm), while for the rollers, it is in the range 90–120 mHz depending again on kiln rotations. It is also possible to compare all harmonics of the kiln at the same time as shown in Fig. 7.17. This creates a better impression of behavior of separate stations because there is difference between the stations and rollers.

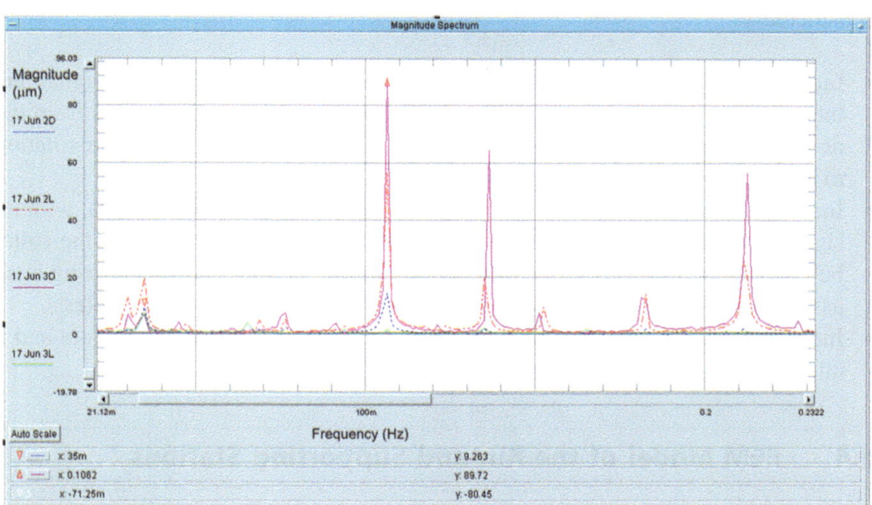

Fig. 7.17 Comparison of FFT graphs of all rollers at the kiln in the same time (*Legend* 2D-*right-hand side* 2nd station, 2L-*left-hand side* 2nd station; 3D-*right-hand side* 3rd station, 3L-*left-hand side* 3rd station)

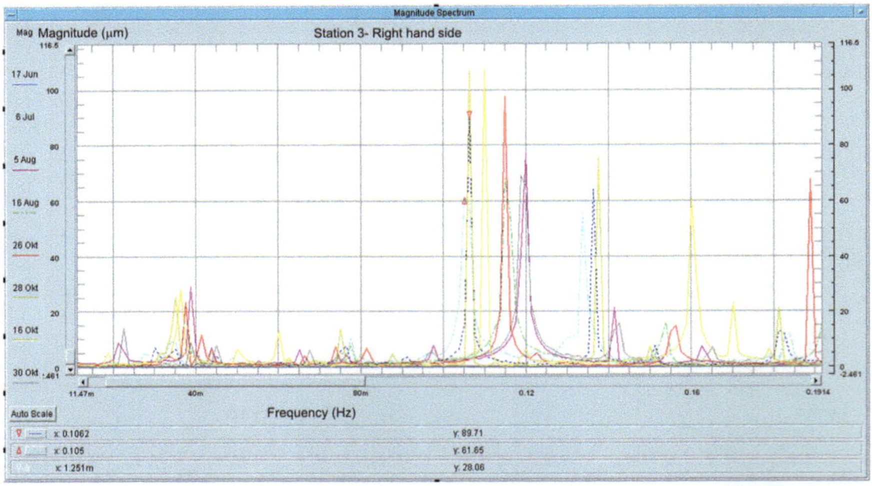

Fig. 7.18 Comparison of FFT harmonics of one roller (3rd station, *Right-hand-side* roller) at different dates

Another useful tool is to compare the baseline measurements with any other measured set. It is possible also to compare any other two measurements to see the direction of change in the condition. In Fig. 7.18, an example is shown where measured sets from several different dates are shown for the station no.3 right-hand-side roller. Reference set is included (condition as described previously).

When analyzing the separate FFT graphs the following principles should be considered:

- Increase in the magnitude of the harmonic representing kiln (KH), is a signal that there is something happening with the kiln and it has larger deviations from the normal operation in terms of eccentricity and deflections from regular rotation axis in section corresponding to monitored station.
- Increase in the magnitude of the harmonic representing the roller (RH), without significant change in kiln harmonic, means that there is a change in the roller behavior, such as orbiting due to processes inside the roller journal bearing bush assembly; or less probably a change in the carrying surface of the roller.
- Increase in both means there is very large increase in the tyre radial run out and it affects the roller significantly as well.

7.4 FEM Model of the Kin and Supporting Stations

To determine limits for alarms of when to undertake certain corrective actions, the rotary kiln and the supporting stations were modeled with FEM using NX-I-deas system. Of course, this model can be used for various simulations of exploitation

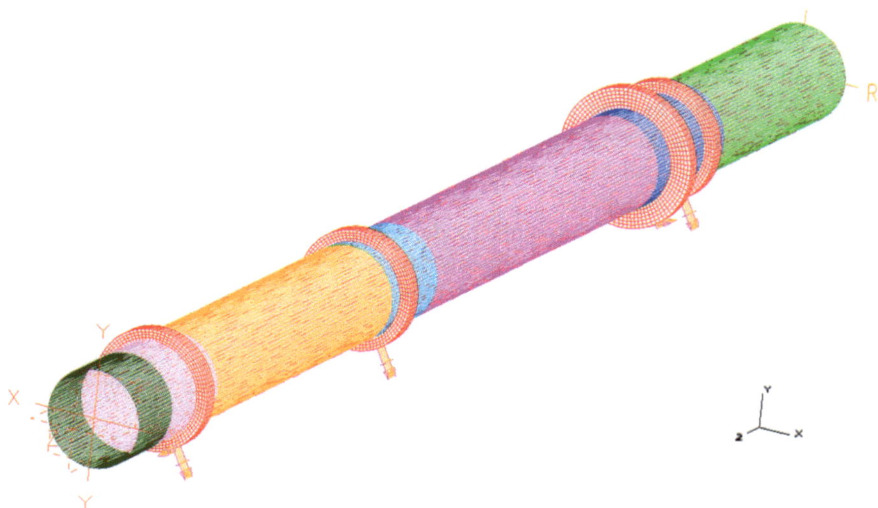

Fig. 7.19 FEM model of the kiln supported on three supporting stations and a girth gear

terms and analyze changes in the equipment condition. There were two FEM models used in the calculations for two purposes:

1. Model of the entire kiln, which includes kiln shell and tyres (Figs. 7.19 and 7.20). This model consisted of SHELL elements (kiln shell) and SOLID

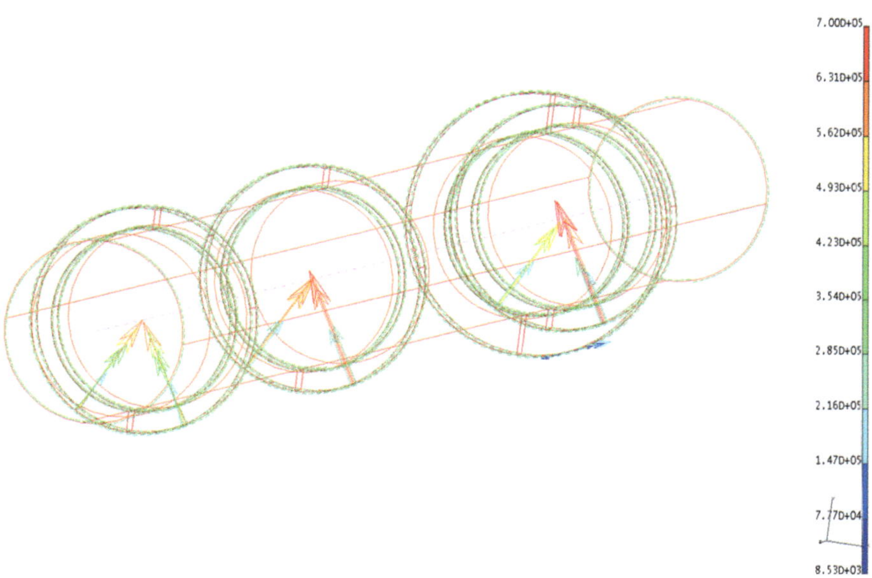

Fig. 7.20 Reactions forces from kiln weight shown as force *arrows* in the kiln supports

elements (tyres and gear rim) with a total of 45,848 elements and 45,144 nodes. Load inside the kiln, consisting of bricks, coatings, and material being burned in the kiln, is represented by MASS elements. The loads include dead loads (kiln own weight, bricks, and coatings) and material inside the kiln. Thermal load coming from the burning operation and melted material is also included in the model. The model of kiln is supported in contact lines of tyres and rollers at three supporting stations.

2. Model of supporting point—this model consisted of SHELL elements (supporting beams) and SOLID elements (tyre, roller, bearing, and its housing) (Fig. 7.21) with a total of 354,453 elements and 543,808 nodes. There are also MASS elements, which represent load inside kiln (bricks, coatings, and material). The model is supported on supporting beams. The loads include dead loads of the kiln (its own weight, bricks, and coatings) and material inside.

The reaction forces (direction perpendicular to roller tyre contact line) for each station obtained by FEM and being a result from kiln uniform load (own weight, bricks, material) of 965t and additional 170t of coatings material located at outlet side distributed over the last 20 m are presented in presented in table below.

Station no.1	Station no.2	Station no.3
4403.2 kN	4406.5 kN	3970.2 kN

The first model describes the operation of the kiln: normal rotation of the kiln with material traveling along its axis from the inlet to the outlet and stresses in the different sections determined by the burning zone as shown in Fig. 7.22.

This model was used to simulate some of the upset conditions of the kiln like the cases of uneven coatings and generation of faults like crank effect as shown in Fig. 7.23.

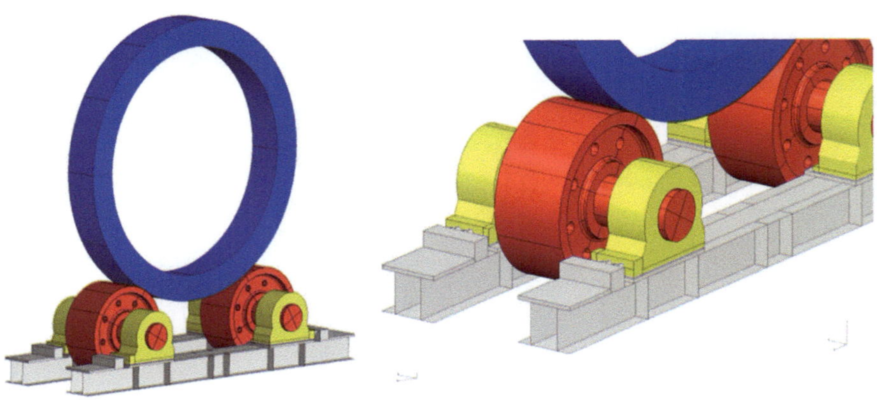

Fig. 7.21 FEM model of the support consisting of tyre, roller, plain bearing, and a housing

Fig. 7.22 Model representing normal operation of the kiln (Huber Mises stress in the kiln structure)

Fig. 7.23 Model representing crank effect in upset condition (Huber Mises stress in the kiln structure)

Fig. 7.24 Huber Mises stress in the roller assembly

The second model describes the supporting system at the supporting stations, but it is worked out in detail the impacts of the loads to each component, i.e., the roller body, the journal to the plain bearing, and housing. The obtained loads from FEM calculations of the entire kiln were applied to the detailed model of the supporting station of the kiln including tyre, roller, bearing housing, and steel frame. Results of calculations of such FEM model are presented in Figs. 7.24, 7.25, 7.26, 7.27, and 7.28. They represent stress level and deflections of the station in direction of measurement of the eddy current probes locations.

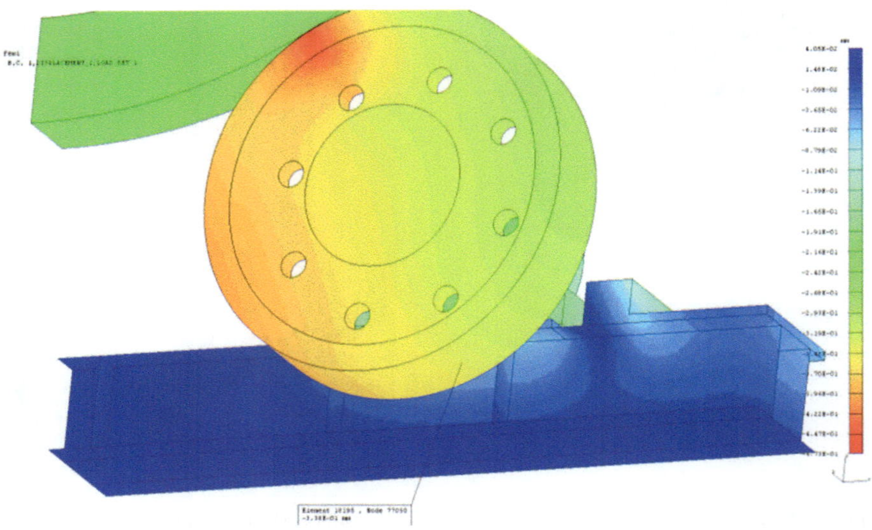

Fig. 7.25 Roller deflection along probe axis (in the probe location) = −0.335961 mm

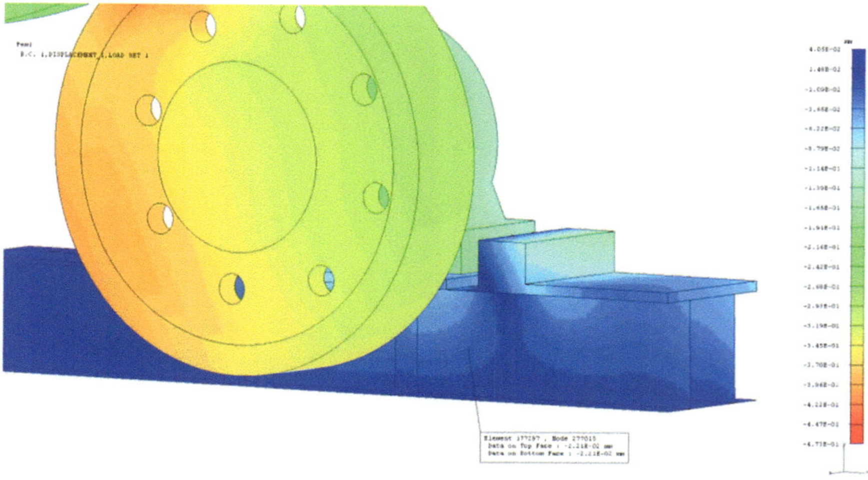

Fig. 7.26 Supporting construction deflection along probe axis (probe location) = −0.0221 mm

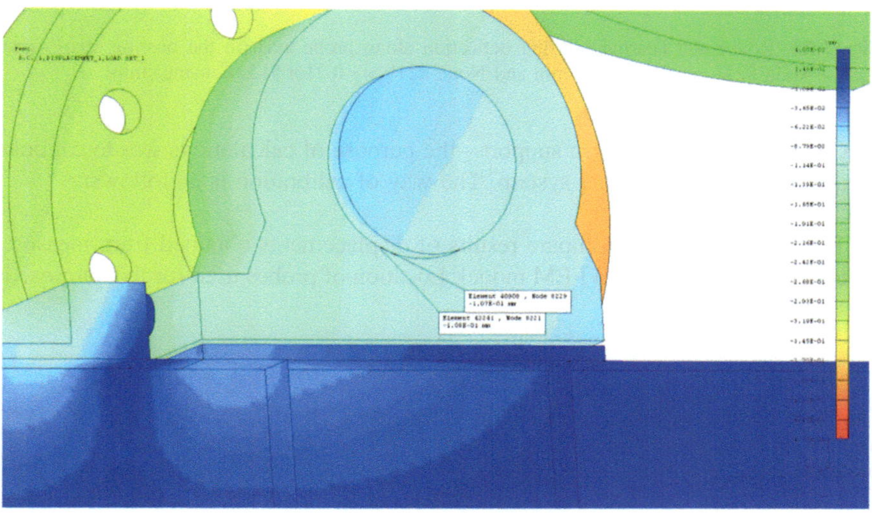

Fig. 7.27 Roller and bearing housing deflection along probe axis (in the bearing area—*outer side*)—Node 8229 = −0.1066856 mm and Node 8221 = −0.1075905 (housing) mm

The FEM calculations were done in two steps:

- Calculations of the model of the entire kiln—the purpose of calculations was to obtain loads transferred through each station for considered internal (material and bricks) condition.

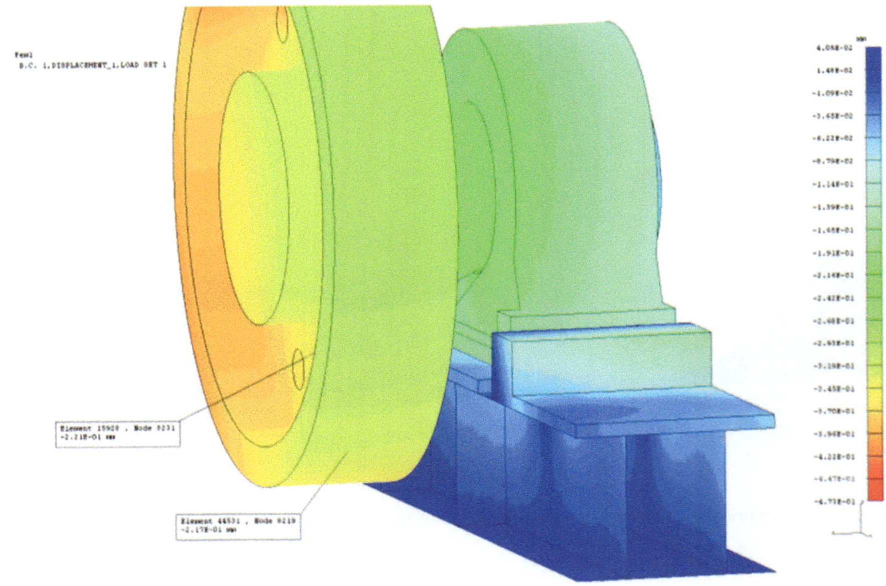

Fig. 7.28 Roller and bearing housing deflection along probe axis (in the bearing area—*inner side*)—Node 8231 = −0.2208863 mm and Node 8219 = −0.2166513 (housing) mm

- Calculations of the station support—the purpose of calculations was to calibrate displacements measuring system. The way of calibration is as follows:

The first step was to compare results of displacements obtained from measurements on the kiln and from FEM model. Location of probes used for this purpose is presented in Fig. 7.29.

Fig. 7.29 Location of probes used for checking of the FEM model on-site

Measurements show that displacement signals taken on the roller and bearing housing are in phase (Fig. 7.30) and the difference between both signals equals:

$$\Delta \text{def}_{,\text{exp}} = 0.05\,\text{mm}$$

Using linear approximation to obtain deflection of the bearing housing at the probe location (680 mm from the bearing) deflection of this point was calculated as presented in Fig. 7.31.

At the figure deflection of the probe locations at the bearing housing obtained from FEM is marked and it equals:

$$\text{def\_h}_{,\text{FEM}} = 0.28\,\text{mm}$$

While deflection of the roller obtained from FEM equals:

$$\text{def\_r}_{\text{FEM}} = 0.335\ \text{mm}$$

The difference between both deflections equals:

$$\Delta \text{def}_{,\text{FEM}} = 0.055\ \text{mm}$$

while difference between corresponding deflections, but estimated via signals obtained from measurements ($\text{def\_h}_{,\text{MEAS}} = 0.08$ mm; $\text{def\_r}_{\text{MEAS}} = 0.13$ mm) equals:

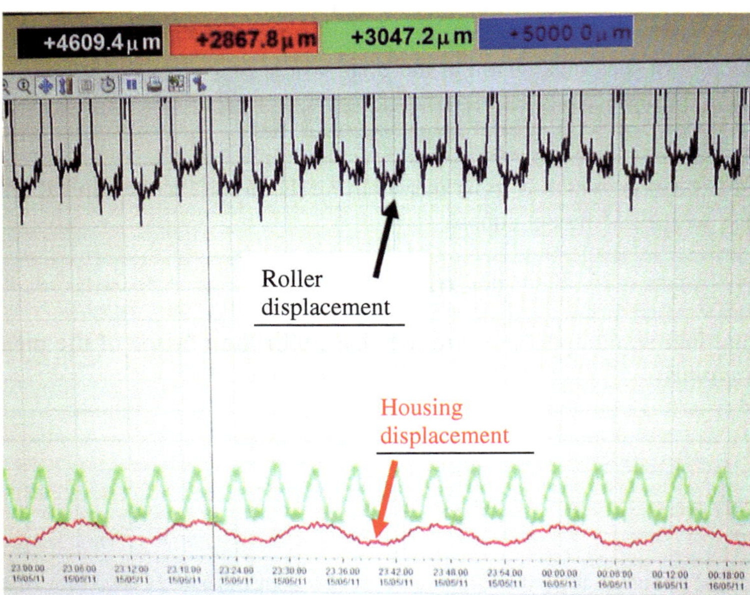

Fig. 7.30 Graph of deflections measured on-site

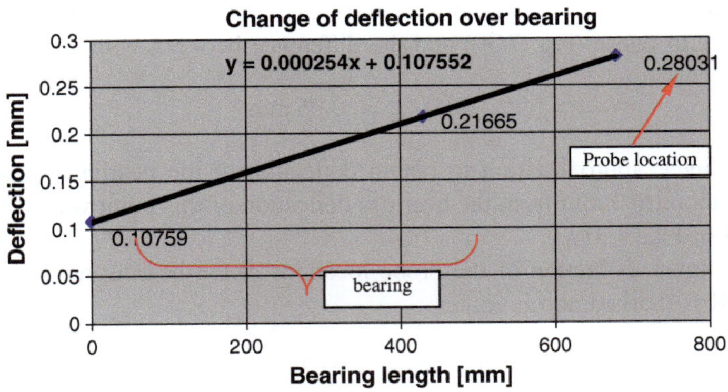

Fig. 7.31 Change of deflection over bearing length

$$\Delta def_{,exp} = 0.05 \text{ mm}$$

There is good correlation between FEM model and experimental measurements.

The second step is actually to estimate the calibration factor. Considering above the FEM model is appropriate for calibration of the displacements measuring system.

From the FEM model deflection of the roller at the station no.2 under the load of $F_2 = 2{,}203$ kN equals:

$$def\_r_{FEM} = 0.335 \text{ mm}$$

Deflection of the construction at the point where probe is fixed is:

$$def\_p_{FEM} = 0.022 \text{ mm}$$

Total deflection, which is to be measured on the probe for the applied load, but estimated with FEM model equals:

$$def\_r_{FEM} = 0.357 \text{ mm}$$

Taking into account the above results, the **calibration factor of the measured system equals:**

$$kf = 6{,}171 \text{ kN/mm}$$

For the maximum allowed loads on such type of bearing of about 21.000 kN/m^2, the limits for the maximum allowed load can be set based on the deflection measured on the roller.

7.5 Examples of Early Detection of Faults Generation

Examples of detecting some faults using previously described measuring and analysis system are given for using the system for online and offline estimations.

7.5.1 Online Estimation of Kiln Condition

Online estimation of the kiln condition and direct possibility to estimate any phenomena is of crucial importance as very often this is exactly the right starting moment to react. The tools to be used here include:

1. Change of the deflection span over a roller revolution ($S_{\text{pk-pk R}}$)
2. Change of the deflection span over a kiln revolution ($S_{\text{pk-pk K}}$)
3. Offset of the deflection span (movement of S_{MIN} value to larger or smaller values)
4. Appearance of a sudden step value change.

Any of these four changes shows that the roller has changed its position significantly in relation to normal operation. It could be overall orbiting of the roller, or it could be simple offset of the roller in its bearing due to thermal changes or kiln axial travel, but in the worst case, it could also be a change in the assembly. In some cases, it could be a damage of the carrying surface, but usually these changes do not happen so fast unless a part chips off from the roller body. In general, when dealing only with online estimation and when some corrective actions are required, it is recommendable to perform detailed visual control of the corresponding station, compare the values of the kiln shell scanner, or do manual infrared scanning of the temperature.

7.5.1.1 Stoppage of Supporting Roller

Stoppage of a supporting roller while the kiln is rotating at normal operation represents an effect of a crank. It shows that crank is already significant (it is in most cases temporary crank) and the tyre is lifted up from the roller (see Fig. 7.32) losing the contact totally from the roller. Considering that the roller is rotating due to friction force posed by the tyre, it stops rotating in its bearing until new force is applied. In some cases, this new action of the force is accompanied with an impact on the roller body, thus posing significantly increased dynamical overloading of the roller journals. In the worst cases, it can end with cracks in the journal corners, wear, or other damages in the soft metal of the bearing bush and also damages at the roller and tyre carrying surface.

This fault can be easily neglected as the only way to validate stopped roller is by visual control and roller may be stopped for only a small section of revolution, thus being not so evident. In today's industry, the plant craft is limited and with one CCR (central control room), very often no operators or maintenance staff is visually

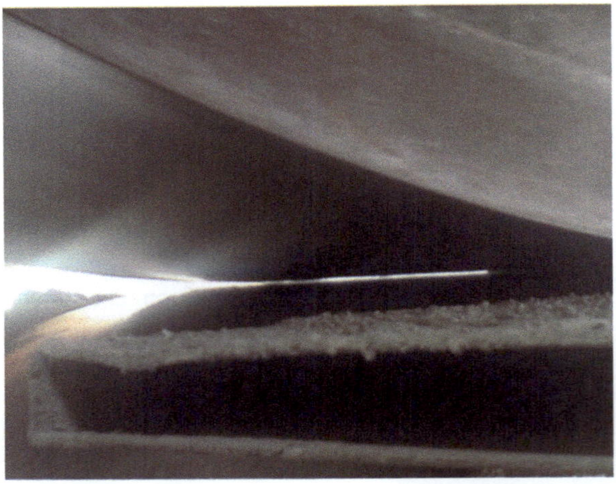

Fig. 7.32 Lost contact between a tyre and asupporting roller

controlling the kiln, especially during second and third shift. Therefore, the situation may run for day(s) even and be recognized when cracks already there. Estimation of this fault in the example was made based on deflection that becomes constant over some short period and the increment of changes is close to zero. It is actually the time period when roller is stopped and the deflection measured receives the same values over several readings. In some cases when stoppage is very short compared to measurement system sampling time, it can be recognized by changing the shape of the deflection graph in the CR where flattened areas appear.

One of such cases has been faced and recorded at the exemplary kiln on its station no.2, left-hand-side roller. The online tools have shown change in the shape of the deflection graph (curve shape) as given in Fig. 7.33. Flattened tops are marked and they are easy to be compared and difference to be noticed with the previous "normal" shape of the deflection changing curve.

The verification was done by physical evidence on-site where really a condition of very poor contact over the entire revolution of the kiln to the rollers on station no.2 was noticed and even more, in a short time interval of a few seconds, the roller at first slowed down and then stopped for a very short period. Photos evidencing this case are given in Fig. 7.34.

The consequences of this case were stoppage of the kiln of about 8 h (lost production of about 15.000 €), shutting the kiln coatings inside with an industrial gun, and special thermal treatment by experienced operator with maintenance supervision to slowly straighten the kiln. Roller journal was later checked against cracks and none was found at that time. However, in 2 months, a crack in the tyre appeared and had to be welded.

7.5.1.2 Kiln Crank (Detection with On-Line Tools)

Detection of the kiln crank is based on the effects of the crank itself on the kiln components.

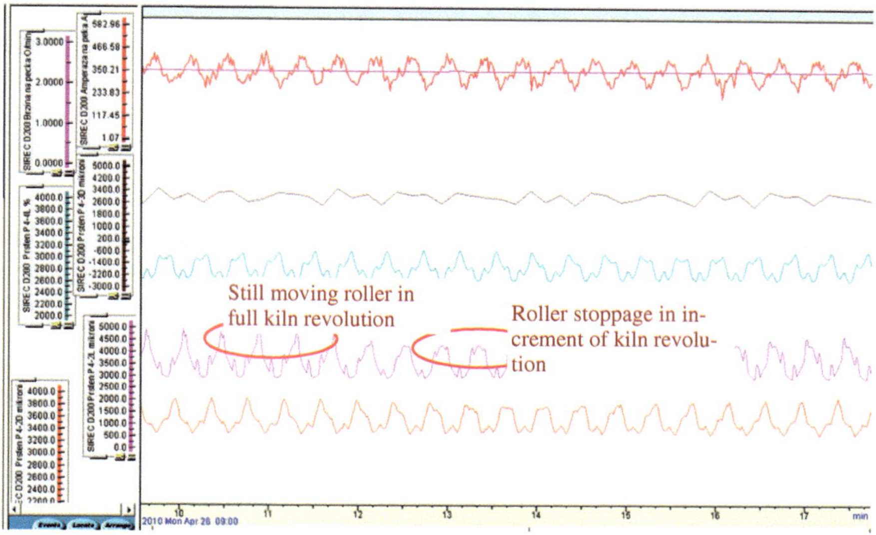

Fig. 7.33 Case of roller stopped over small increment in kiln revolution shown by flattened tops of the deflection changing curve

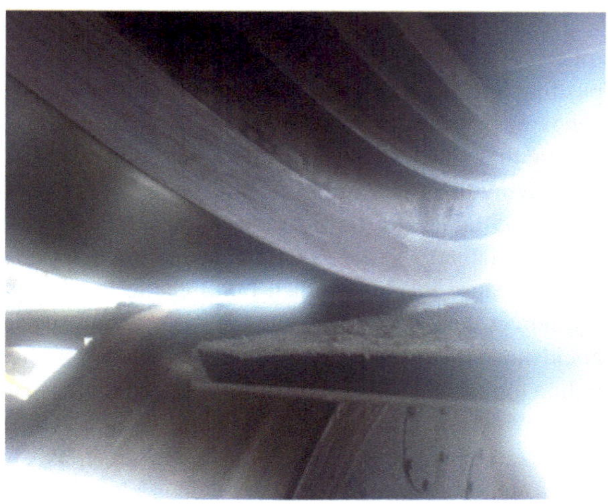

Fig. 7.34 Poor contact resulting in short stoppage of the supporting roller (*left-hand-side* roller at station no.2)

Very precise description of how and where crank can appear was given by Z. Krystoxczyk in his work: "Geometry measurements of kiln shell in dynamic condition" [9] where it is given that there can be two basic types of crank by the place they occur:

- crank due to eccentricity of the shell between tyres,
- hidden crank (mechanical or thermal) when crank appears due to deformations and eccentricity in the supporting stations. This is an especially dangerous crank as at the supporting station, there is no possibility of these deformations to result in geometric eccentricity, but increased force raises acting on the supporting rollers that changes their deflection.

Especially dangerous is the thermal hidden crank that can appear at the supporting station rather suddenly resulting from fallen coatings due to process reasons and even a temperature difference in one circumference of 80 °C can cause significant deflection of roller axes.

With the system presented in this book, tools were developed that are in line with the physical consequences of crank presence in the kiln. These tools consist of:

- Visual online representation of the roller radial deflection in time domain;
- Analysis of deflection by filtering the influence of the kiln rotation;
- Analysis of the magnitudes by FFT harmonics (frequency spectrum) representing the behavior of the kiln and of the roller separately;
- Analysis of the cyclic nature of the load from the tyre to the one and other roller at the same station in crank condition.

Detection is based on the property of the kiln crank to pose uneven load on the rollers of the same station causing alternative increasing and reducing load, thus pushing and liberating the rollers (as given at the beginning of this chapter). Such behavior is causing increased span of the deflection change over a kiln revolution. So, the first sign of this is actually noticing increased $S_{\text{peak-to-peak}}$ values in the graph and having this trend for a longer time. It is easy to notice visually and is shown in Figs. 7.35 and 7.36. In this case, there was a thermal crank and the situation was lasting relatively longer without any corrective actions. The crank was not very huge to cause loss of contact, but a cyclic loading was present and the final **consequences** were severe damages of the carrying surfaces of both tyre and roller that could not be caused by anything else in such a short time interval. In this case, online tools have shown significant change in the span of the deflection change at a kiln revolution and it could be seen by the deflection changing curve on the visualization window. In Fig. 7.35, the graph is given and separately S_{MIN} and S_{MAX} are marked for their values. It is also evident by the trend that there was progressive increase in the deflection span over the last day and particularly over the last few hours.

Deflection span has been enlarged from a regular value of 0.7–1.5 mm at the left and right-hand sides from 0.3 to 0.5 mm at rollers at station no.2 (visual evidence in the control room as shown in Fig. 7.36). The load was changing as contact area through which load was transferred has been alternatively changing from 70 % down to 20 % in a cycle during every kiln revolution. The **consequences** after 24 h were huge damages on both the carrying surfaces of the tyre and rollers (Fig. 7.37).

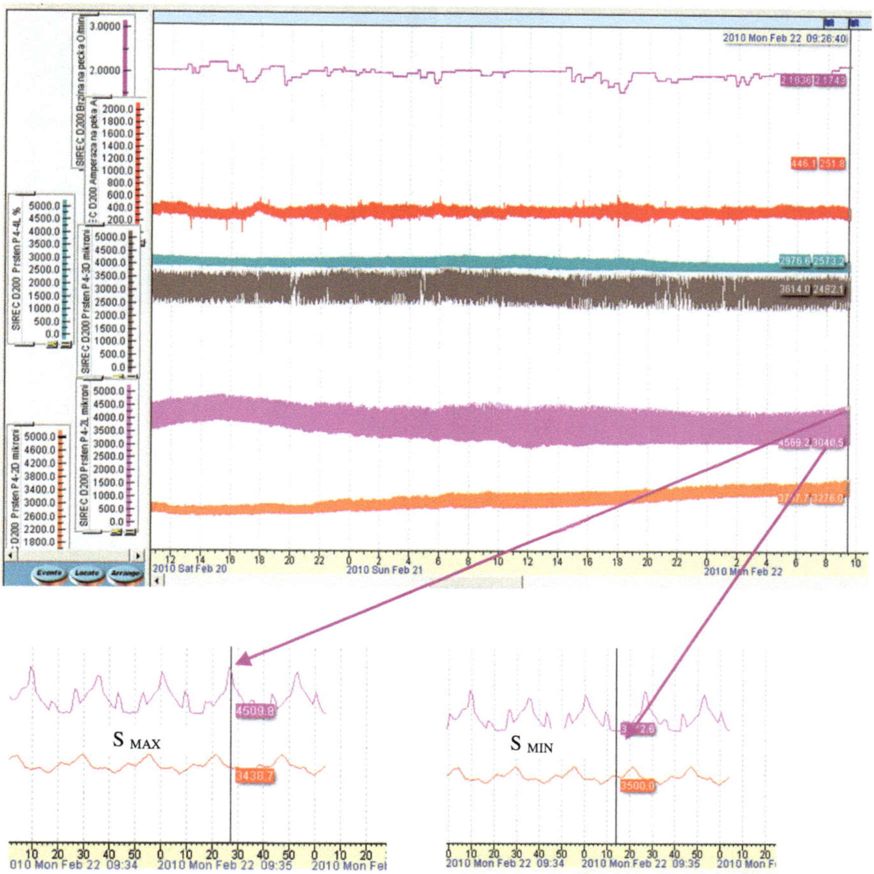

Fig. 7.35 Increase of total deflection of the roller in a kiln revolution pointing to kiln crank

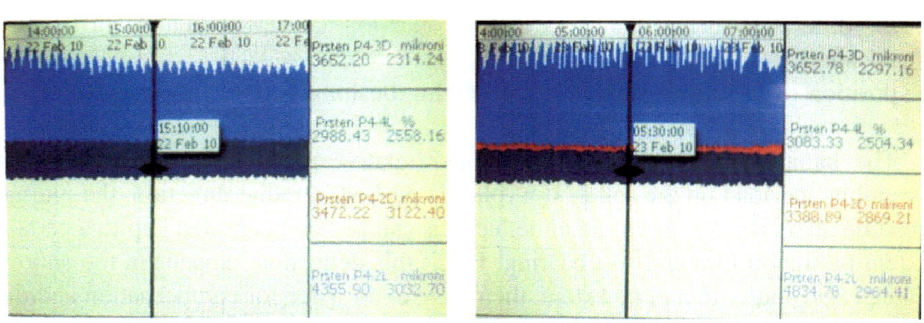

Fig. 7.36 Evidence of rollers deflection increase in CR visualization and recording graph

Fig. 7.37 Damages on the carrying surface of the supporting rollers and tyre due to kiln crank

7.5.1.3 Hot Bearing at Rollers

The online estimation enables also detection of one specific condition when a roller is being moved in its bearing bush in a significantly short time, so that there is not enough time for the bearing soft metal to compensate the new position of the roller journal. Inside the assembly, the oil film is reduced significantly, thus being no more able to sustain the load resulting in change in the lubrication regime from hydrodynamic to boundary. In this case, the reduction of the oil film is the root cause for the hot bearing. The actual oil film in these bearings is determined by[1]:

- the difference in the diameters of the roller journal and journal plain bearing bushing;
- position of the roller journal in the bearing bushing.

The latter is in balance with the roller relative position in the overall kiln assembly and load on the roller. If a roller is pushed in radial direction, the journal position also changes. Such radial deflections can be "compensated" up to a certain extent by the oil film and soft bearing. But if this deflection happens in too short a time period and/or to a large extent, then there is no space for compensation and the

[1] Parameters that are not changing in short time interval under exploitation (as surface roughness, oil quality, assembly properties) have not significant effect here for the described changes.

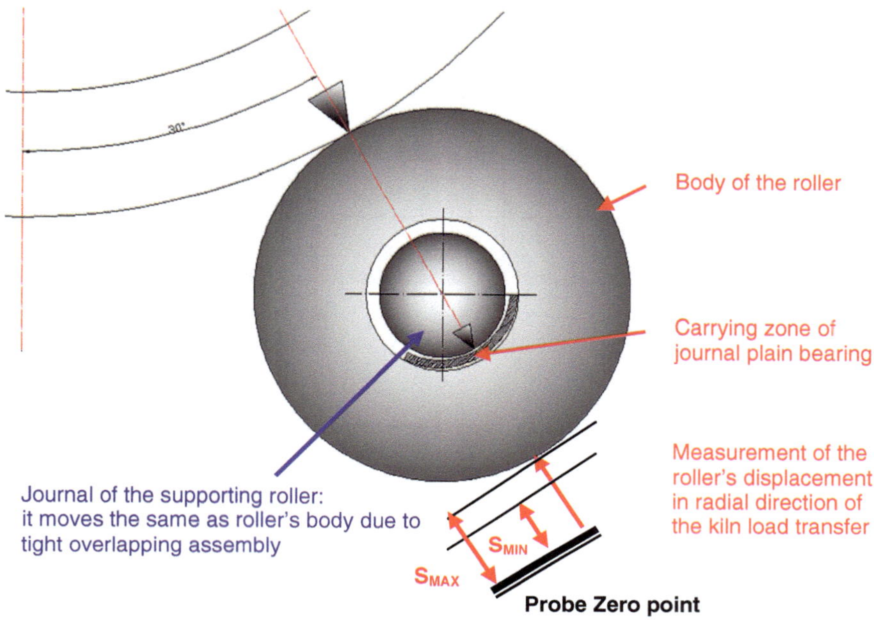

Fig. 7.38 Roller deflection causing reduction of oil film; measurement of deflection relative to external fixed point (Min Oil thickness is 0, 2005 mm for the exemplary supporting rollers under examination)

above described preconditions for hot bearing take place. This proves that for hot bearing, it is important to monitor the distance of the roller journal to the bearing bushing in load transferring direction (Fig. 7.38).

In order to confirm this relation, examinations have been made [1] of the change in the bearing temperature in correlation to the S_{MIN} value of the roller displacement in real conditions under the following terms:

- uniform ambient temperature,
- uniform shell temperature (as the kiln shell is affecting bearing temperature),
- equal kiln capacity (kiln load).

Measured S_{MIN} deflections and corresponding bearing temperatures (following the scheme of Fig. 7.39) are represented with graphs in Figs. 7.40 and 7.41.

All graphs show that bearing temperatures and S_{MIN} values have clear dependence. They are inversely proportional, which is clear because the smaller the S_{MIN} value, the closer the journal is to the bushing in the bearing assembly, and the thinner the oil film becomes the higher the bearing temperature. Trends of decreasing the S_{MIN} values are accompanied with trends for increasing the corresponding bearing temperatures and vice versa.

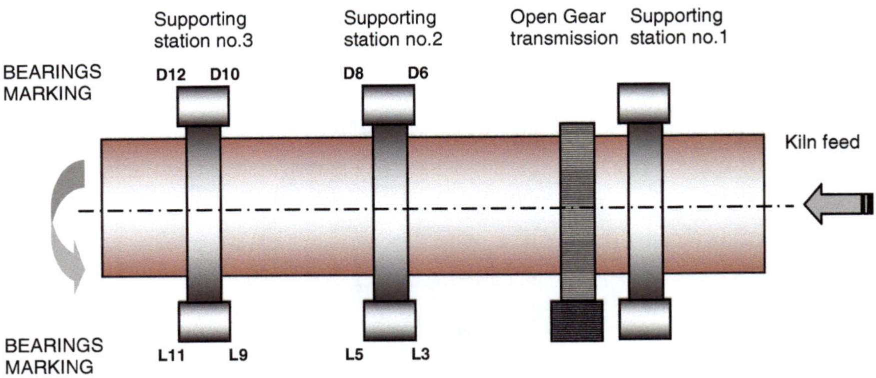

Fig. 7.39 Bearings marking at separate supporting stations

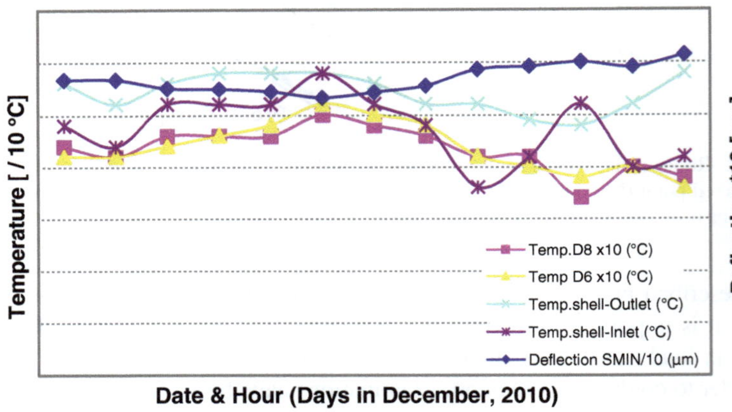

Fig. 7.40 Correlation bearing temperature and MIN value of displacement—st. no.2, *right-hand side*

Obtained data were checked against their correlation and finally quite strong correlation coefficients were obtained ranging from −0.6 up to −0.92 for S_{MIN} to bearings temperature [1].

A hot bearing case was recorded when the temperature was increased to a level defining the bearing as "hot bearing" in little more than 1 h. The case was recorded with the temperature monitoring module of the CR controlling system and is shown in Fig. 7.42 together with all temperatures of the bearings at the corresponding station. Separate temperature trends are marked in different colors and the hot bearing peaks at 13.54 h. In the particular case, the maintenance team was on-site and corrective actions were undertaken to stop further temperature increase. If not successful, the soft metal could melt at 60–70 °C, thus causing kiln stoppage of at least 2 days (if spares available) and causing loss of production of about 190.000 €

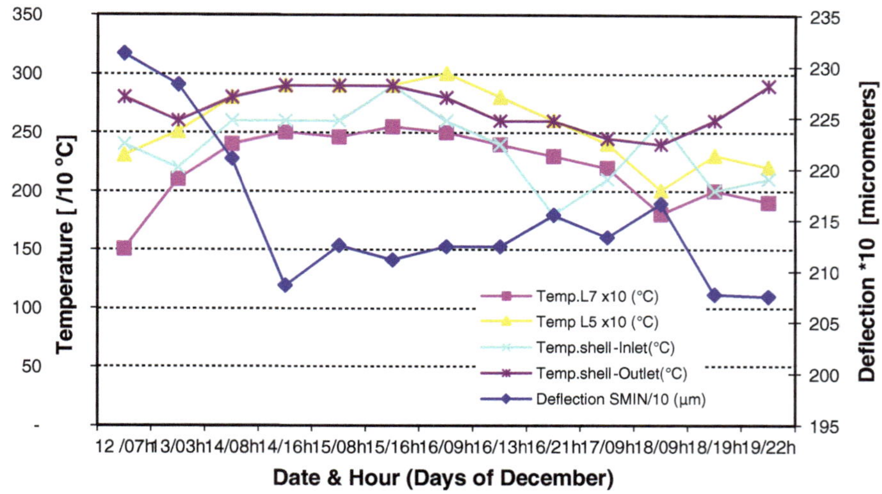

Fig. 7.41 Correlation bearing temperature and MIN value of displacement—st. no.2, *left-hand side*

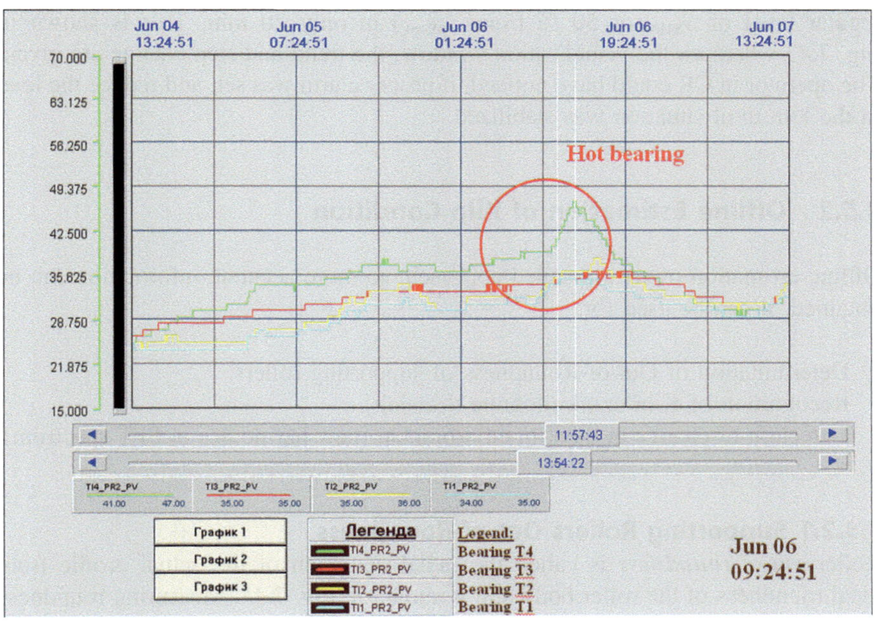

Fig. 7.42 Case of hot bearing at station no.2—*Left-hand side*—uphill—exemplary case

and additional costs for repair including spare parts, cranes, third-party labor of about additional 25.000 €. This case could have been recognized by the step change of the roller S_{MIN} value as it has been changed for about 0.2 mm (about 6 % from

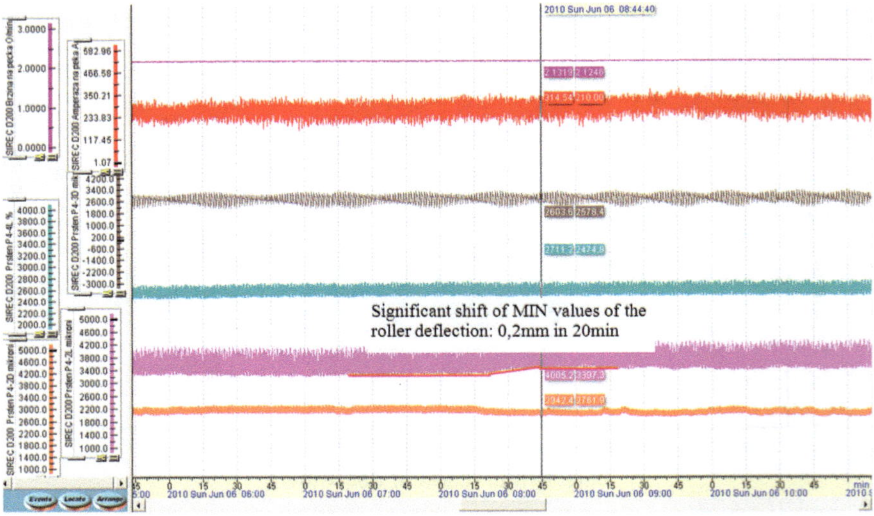

Fig. 7.43 Step change in S_{MIN} resulting in later increase of bearing temperature

regular level of S_{MIN} or 30 % from $S_{pk\text{-}pk}$) in only 20 min. This is shown in Fig. 7.43 where on the visualization window, the trend and step change are given. The operator in CR could have noticed, if proper alarm was set, and reduce the load in the kiln until situation was stabilized.

7.5.2 Offline Estimation of Kiln Condition

Offline estimation offers a detailed approach, so more accurate information can be obtained. It can be used for:

- Determination of Out-of-Roundness of supporting rollers.
- Recognition of Kiln crank (definite decision).
- Detection of change in the kiln environment (new harmonics in FFT spectrum).

7.5.2.1 Supporting Rollers Out-of-Roundness

Roller ***out-of-roundness*** is called the radial deviation of the actual profile from ideal roundness of the roller body and is shown in Fig. 7.44. Measuring roundness requires 360° traces of the measured workpiece. Turntable-type instrument or a stylus-type instrument can be used. For purposes of measuring in real condition on-site, a stylus-type instrument is convenient. It records the distance from the instrument tip to the workpiece that is turning. The out-of-roundness is a consequence of the wearing process.

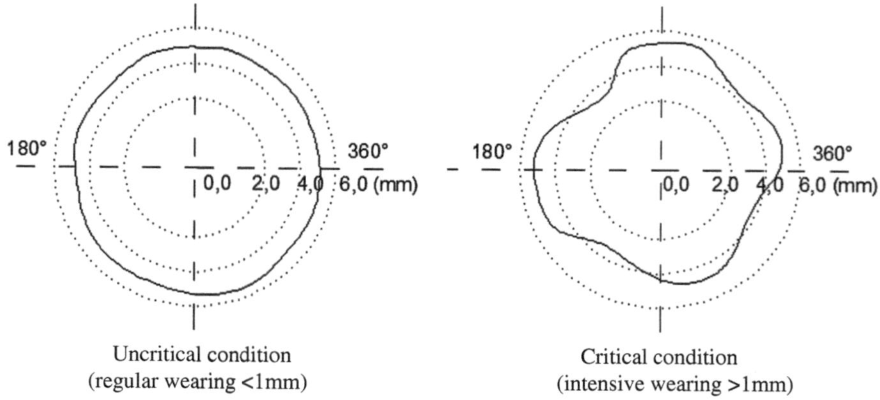

Fig. 7.44 Out-of-roundness condition of kiln roller surface

Actually, wear of tyres and supporting rollers is a critical phenomenon, which is a combined consequence of the other phenomena as kiln crank, uneven load, as well as consequence of the nature of the kiln operation and dusty environment. Since it cannot be avoided, no matter of all is made to monitor and prevent, in the lifetime of these components often is necessary to resurface them. The maximum allowed wear and machining of the tyres is about 5 % of the original diameter [3, 10]. Good practice is rollers to be resurfaced after they are out-of-roundness more than 1 mm. Higher out-of-roundness can cause increased vibration of the corresponding station of the kiln and therefore **monitoring of out-of-roundness** of the roller is of significant importance. This is especially the case if there is continuous ascent and descent several times over a single rotation of the roller (Fig. 7.44). However, until there are no obvious negative effects, the tyres and rollers are usually left to operate. In the final case, grinding has to take place in order to correct the roller roundness.

Kiln rollers are of huge dimension and the accuracy required is of low level (0.01 mm), so measurements in three planes are sufficient and also applied by commercially available measuring companies.

When applying the new proactive monitoring, in order to improve the measuring accuracy and minimize the influence of the tyre radial run out, it was estimated that it would be best to measure when kiln is not in operation (eliminating thermal crank effect) and rotate the kiln at very slow speed to provide more measuring data as the sampling is set to certain finite time (e.g., 1 s).

Such measurements were performed at the exemplary kiln-st.2 and examples are given in Figs. 7.45 and 7.46. Results were also confirmed with results obtained by other appropriate commercially available methods [1, 5, 10–12].

Measurement of the out-of-roundness used in the example was performed at kiln with new bricks and no coatings, so free of any operational influences. The following were the measurement conditions:

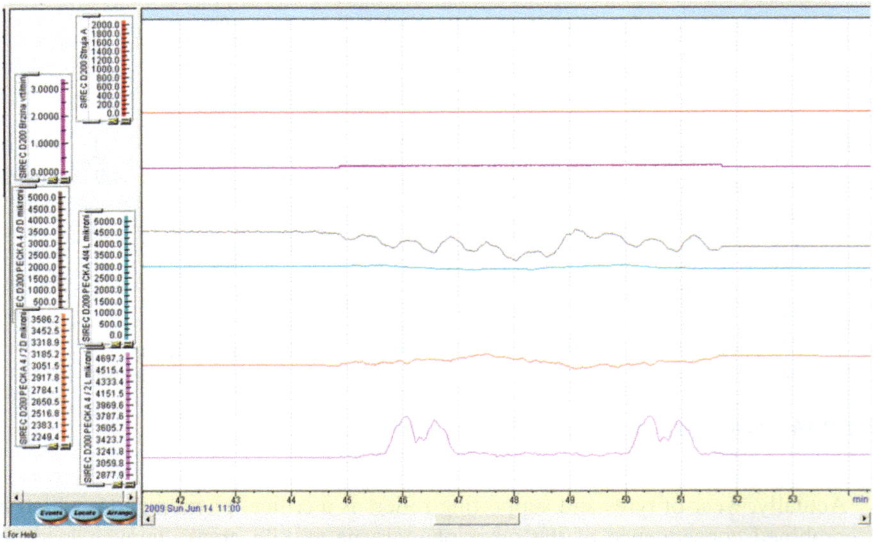

Fig. 7.45 Rotation of the kiln for the purposes of out-of-roundness measurement

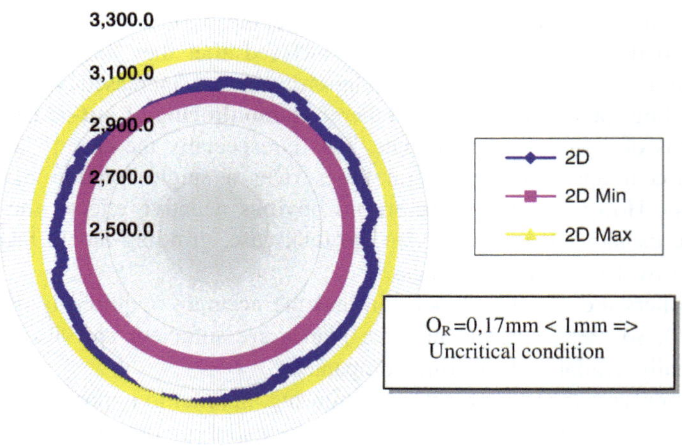

Fig. 7.46 Out-of-roundness graph after mathematical processing of *Right-hand side* St.2

- Kiln rotation while measuring: n = 0.14 rpm
  - 1 kiln rotation durability: 7.14 min or 428 s,
  - 1 roller rotation durability 7.14/3 = 2.4 min or 142.8 s
- Kiln in cold condition without coatings (new bricks installed)
- Max and Min value of a roller diameter were estimated with increased density of sampling points (142 points at 1 s sampling rate, or 11 mm distance between two points at periphery of 1,573 mm).

Measurement was recorded and the graph is shown in Fig. 7.44.

Displacement values for this case show distances from the roller profile to the probe. Measured data were further processed and Orbit plots were made to represent the out-of-roundness (O_R) of each roller. Here, only one roller plot is given as example in Fig. 7.46.

7.5.2.2 Kiln Crank

Two examples are given describing two different kiln conditions: normal and condition of crank. In addition, first analyses were carried out in parallel with external expertise measurement company in kiln crank-free condition and results were confirmed [1].

Kiln Crank-Free Condition

At startup of a mechanical crank-free kiln, when there are still no uneven coatings in the kiln, especially after new lining has been installed and shell temperature is approximately even, the kiln during rotation affects rollers uniformly (relatively equal to their position). Parameters representing this condition are given by the rollers deflection visualization (Fig. 7.47), kiln shell scanner representation (Fig. 7.48), FFT analysis (Figs. 7.49 and 7.50), and fitted SINE function in (Figs. 7.51 and 7.52). **These measurements represent also the baseline measurement** taken at the same time.

Kiln Crank Condition

A case of kiln thermal crank was analyzed at the exemplary kiln. The crank origin was the significantly uneven coatings causing also significant shell temperature

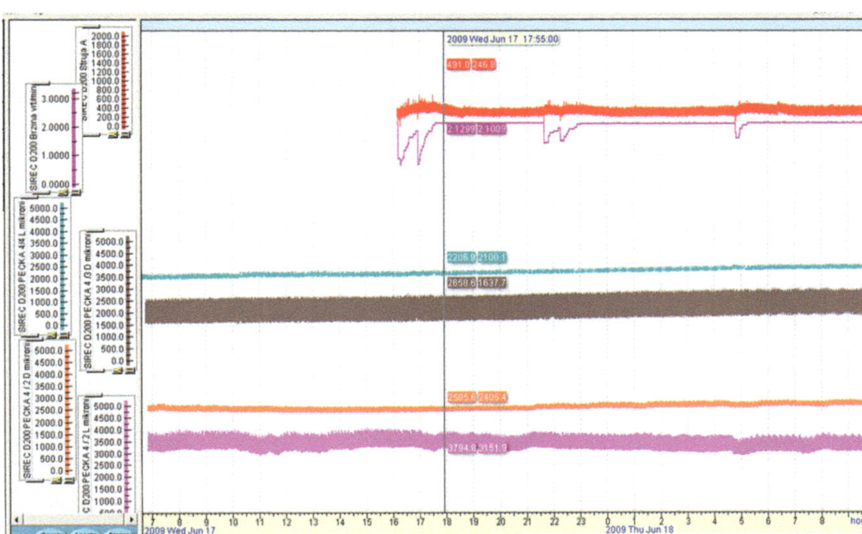

Fig. 7.47 Rollers deflection *right* after kiln startup

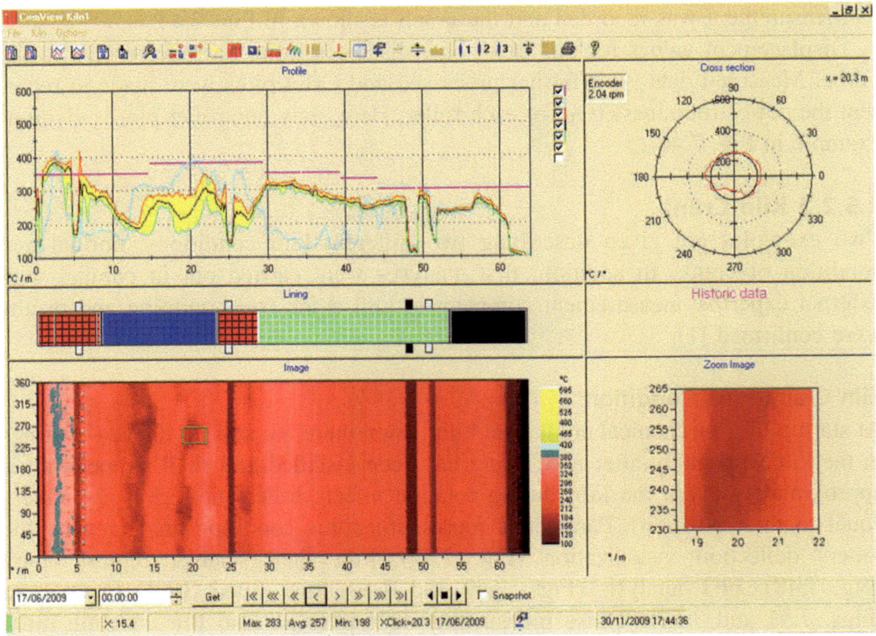

Fig. 7.48 Kiln shell in crank-free condition represented via shell scanner

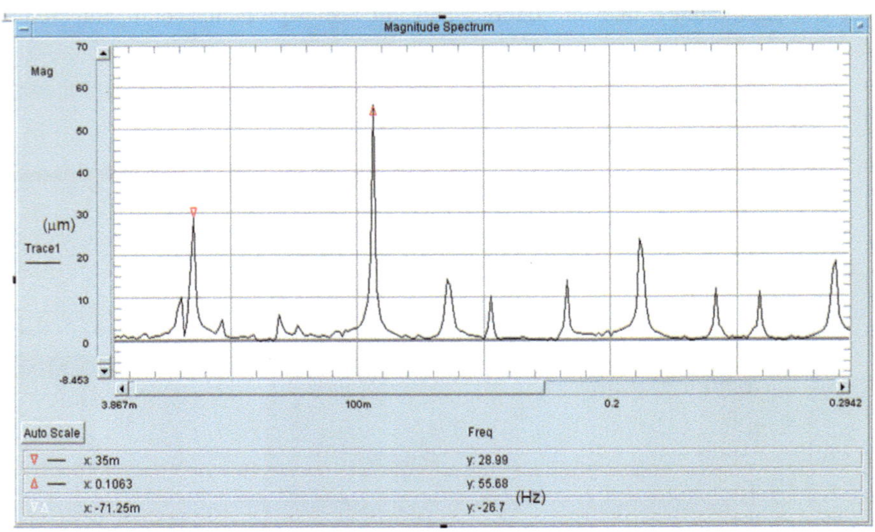

Fig. 7.49 FFT of roller deflection at crank-free condition—st. no.2—*Left-hand side* (Startup)

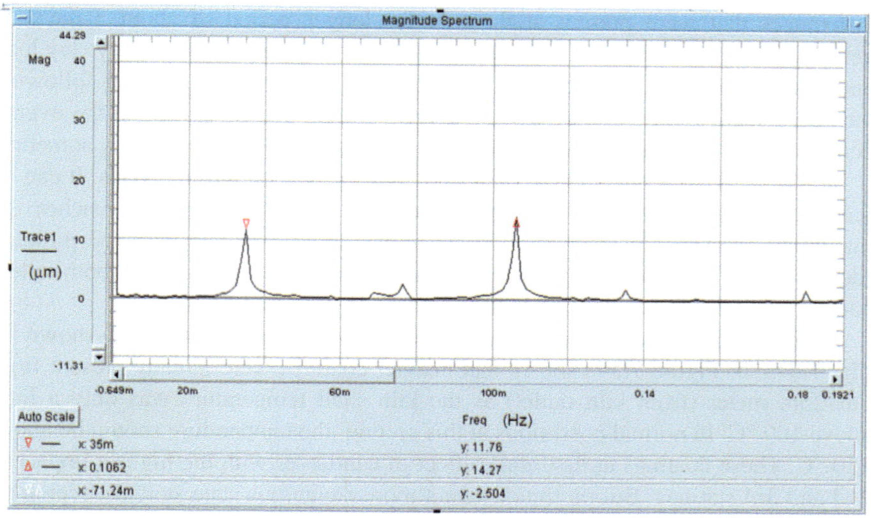

Fig. 7.50 FFT of roller displacement at crank-free condition at station no.2—*Right-hand side* (Startup)

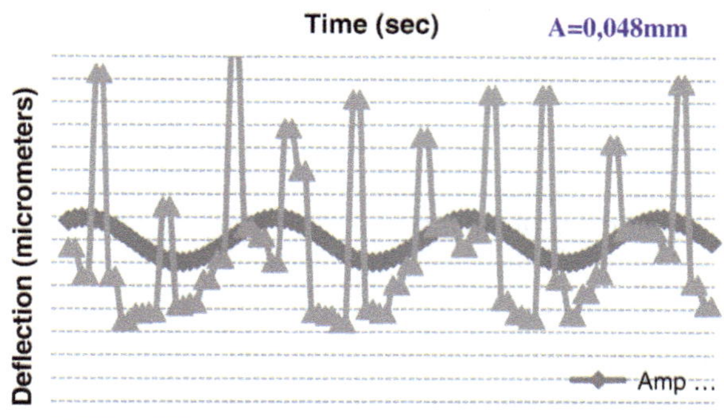

Fig. 7.51 Effect of the kiln (SINE function) on roller deflection in crank-free condition (Station no.2—*Left-hand side*)

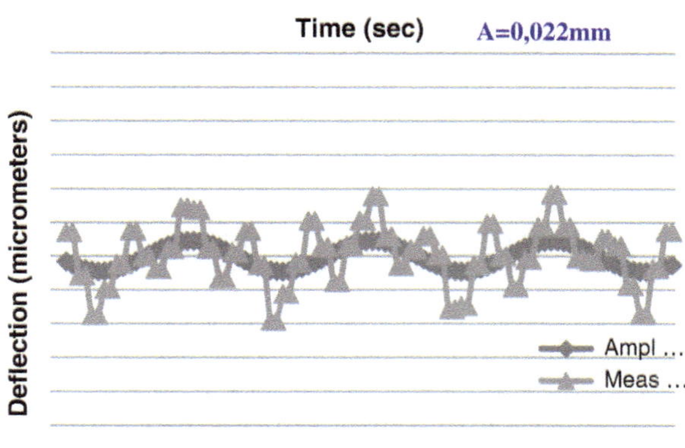

Fig. 7.52 Effect of the kiln (SINE function) on roller deflection (station no.2—*Right-hand side*)

differences that were present at the kiln for longer period of about 3 days. In general, uneven coatings are a frequent case in kiln operation, but if they stay long and at some combination, they can cause huge crank. However, operators following only the scanner cannot be aware of the crank extent and its effects on the overall condition even if contact is lost, unless the latter is visually evidenced by someone.

In Fig. 7.53, roller deflection values are shown for a period of 7 days. It can be recognized that displacement increases starting from April 24, but reaches the maximum on April 26 when the kiln was even stopped to eliminate the problem (details can be found in [1–3]. The crank was formed at station no.2 and both rollers had significantly larger deflection than at normal condition.

State of the uneven coatings and their change through time is shown in Fig. 7.54a, b. Figure 7.54a shows that heavier coatings were already present from 6th–25th meter (from kiln outlet) as the kiln shell temperature was only a little above 100 °C. In normal conditions at this section, the temperature should be above 250 °C. These coatings at first make effect of cantilever with the highest impact to 2nd and 3rd stations. But on top of this, a more dangerous state started developing when some of the coatings started falling down and temperature of the shell between the 15th and 20th meter started increasing above 300 °C, almost to 350 °C. It caused shell temperature difference of more than 200 °C. It is obvious that a cylinder made of metal with such temperature difference (Fig. 7.55) will deform and cause crank of the cylinder. Eccentricity caused is also shown in the upper right-hand-side angle of every scanner image (Fig. 7.54) titled "Cross section."

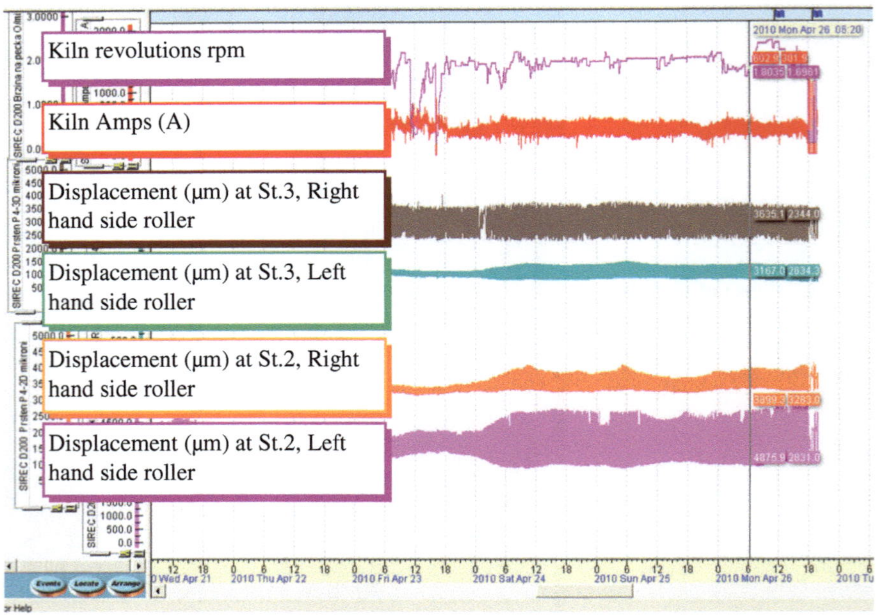

Fig. 7.53 Kiln crank reflected at supporting roller at station no.2 on April 26

(a)

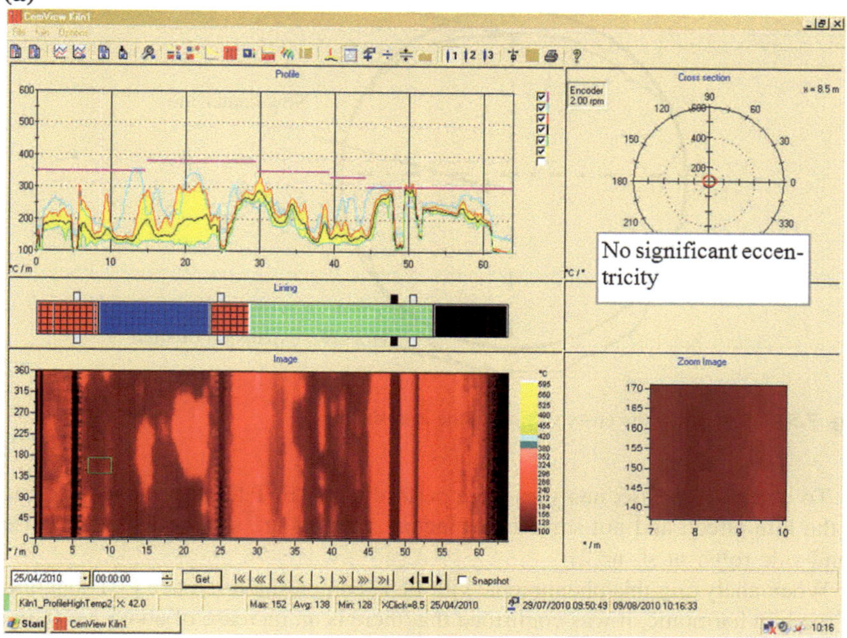

(b)

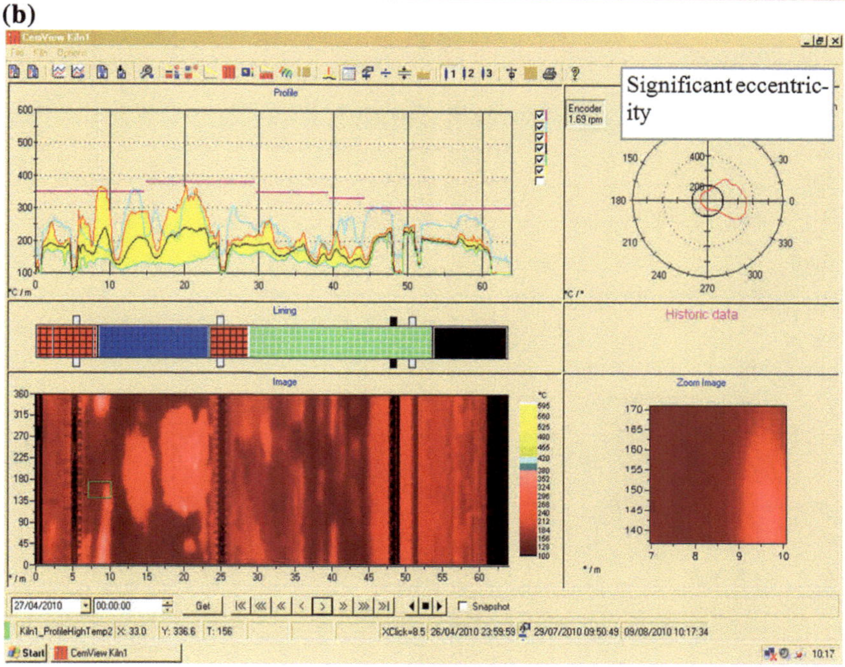

Fig. 7.54 Crank development—Kiln shell condition represented by the shell scanner. **a** Presence of heavy coatings and start of their falling down process. **b** Presence of heavy coatings and extensive falling down process

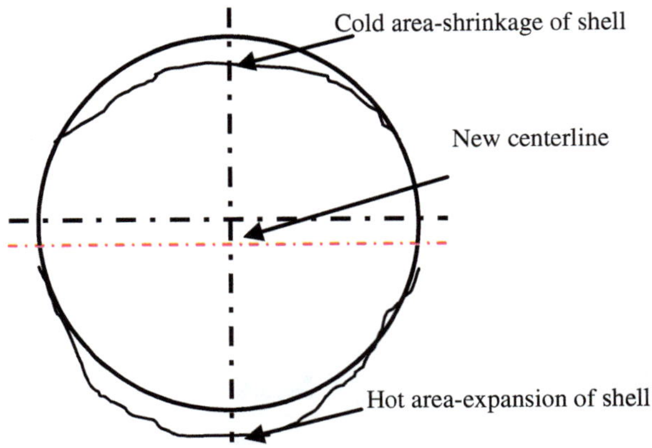

Fig. 7.55 Changes in the cross section of the kiln shell

To confirm the case, analyses were performed by FFT to detect if the real cause is the kiln effect, and not some other factors. Results are shown in Fig. 7.56 (left-hand-side roller at st. no.2)

When analyzing this phenomena with FFT, comparing reference FFT magnitude of the kiln harmonic, it was confirmed that there is an increase of about 8–10 times, while at the roller harmonic, the magnitude was almost not changed (see Table 7.1).

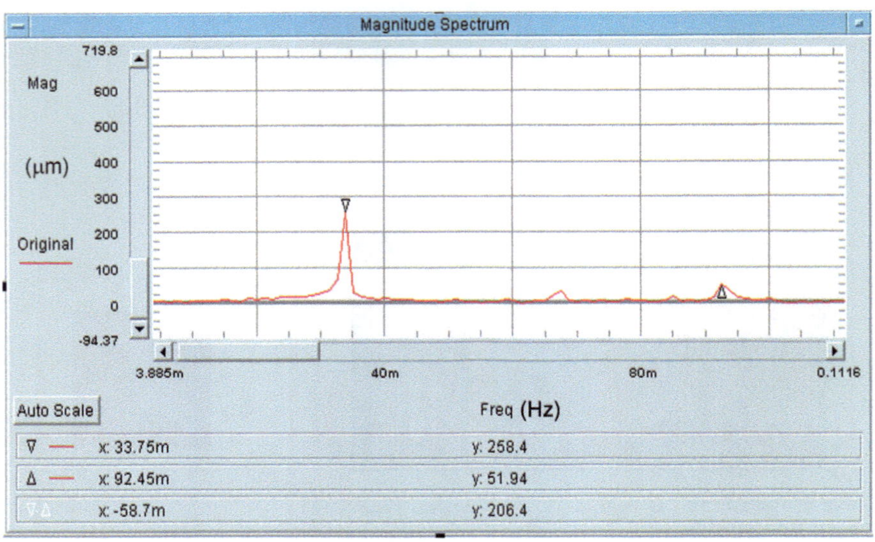

Fig. 7.56 Magnitude of kiln harmonic in FFT is 10× from the corresponding magnitude when kiln is in normal condition (see Fig. 7.49)—*Left-hand side* 2L

Table 7.1 Comparison of harmonics magnitudes at reference and crank condition

St.2-Right hand side	Reference	Crank	St.2-Left hand side	Reference	Crank
Kiln	**11.76**	**87.58**	Kiln	**28.101**	**258.4**
Roller	**14.27**	**15.44**	Roller	**55.68**	**51.94**

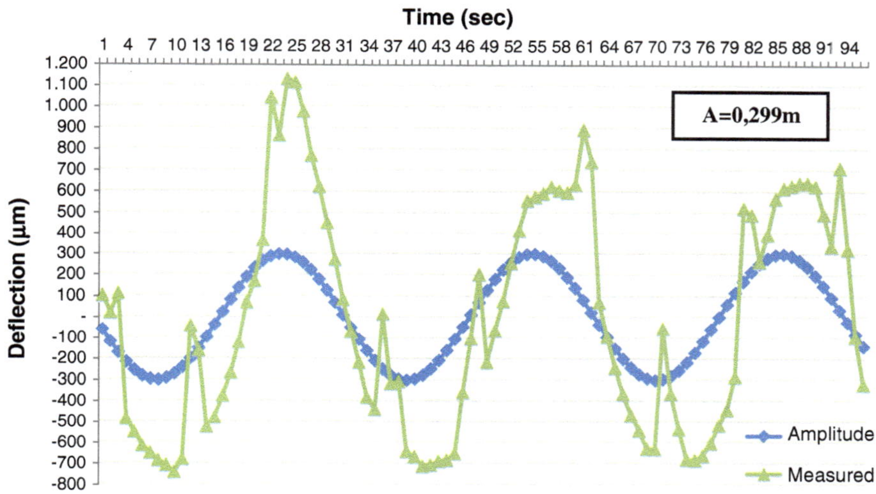

Fig. 7.57 Effect of the kiln (SINE function) on roller deflection (station no.2—*Left-hand side*)—Crank

The same condition (same sampling set) was analyzed by fitting SINE function (see Fig. 7.57). Obtained filtered roller "axial sag" for station no.2 was 0.258 mm, which is larger than 0.2 mm, which following the rules [3], it is recognized as crank condition.

One more method has been used: analysis of the **cyclic load on the two rollers** at the same station based on the other crank effect (causing cyclic load at rollers). In Fig. 7.58, the graphs show that there is phase difference in the peaks of SINE function corresponding to the phase difference in the two rollers placed at the same station. Also, the forwarding graph is the right-hand side rollers that actually come first into contact with the tyre (tyre and kiln has CCW direction looking from kiln outlet to inlet).

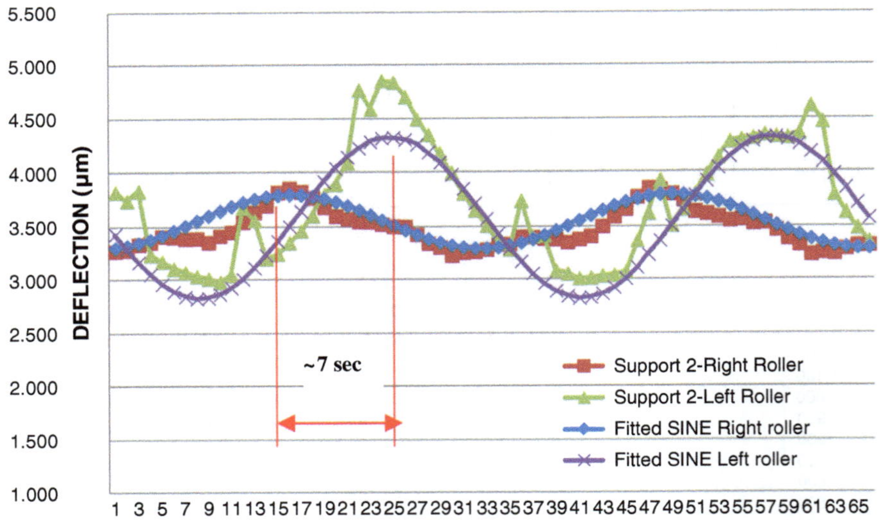

Fig. 7.58 Cyclic load on the two rollers expressed via phase difference in the two rollers deflections at the same station

7.6 Proactive Actions for Early Corrections

The purpose of proactive actions for correction is to provide on-time activities to eliminate or at least bring down to minimum the existence of **the root causes** of certain equipment potential failures, thus reducing their effects and restoring the reliable equipment condition. There can be many different corrective actions that are governed by the nature of the root cause itself, as well as the possible way to influence the equipment.

The **first step** in determining the corrective actions is when carrying out the FMECA for corresponding equipment, to **determine accurately the root causes** of the failures.

The **second step** would be to analyze the **nature of these causes** and determine all **factors influencing** the causes that could boost their progress or factors that could decrease their development and influence.

The **third step** is to predetermine the **possible actions** to be undertaken **in short time** and without special costs and time as the first approach. Of course, if no such possibility, **then regular corrective actions** with all appropriate means should be determined.

The **fourth and last step is to set alarming levels** for the measured parameters **when to take corrective actions** and when to do something more progressive (as stopping equipment operation if no positive effects from previously undertaken corrective actions). Proactive actions for the main kiln failure modes that were analyzed would be the following:

7.6.1 Proactive Actions for Roller Out-of-Roundness

Out-of-roundness as previously described is a phenomena when kiln roller is losing the round shape and thus causing vibrations of the supporting stations affecting overall related equipment. Proactive actions are shown in Table 7.2.

7.6.2 Proactive Actions for Hot Bearing

Hot bearing is a phenomenon when bearing temperature rises significantly in a short time faster than 5 °C in 1 h and approaches limiting values of 45 °C for Al-based alloy or 60 °C for Cu–Sn-based alloys. Following the results obtained with FEM model for the regular conditions, deflection of the housing of the exemplary roller bearing is in the range of 100–200 μm, while roller deflection is about 300 μm. The gap in the journal plain bearing is about 205 μm. In normal conditions, it gives about 100 μm spare space before having metal-to-metal contact.

Based on this, proactive actions are shown in Table 7.3.

Table 7.2 Proactive actions against rollers out-of-roundness

Failure mode	Out-of-roundness
Root causes	Extensive friction forces, cyclic load, overpressure
1st Stage-Immediate actions	Check and correct contact pattern, check and correct roller alignment Keep clean the station surrounding (by vacuum regularly)
2nd Stage-actions	Resurfacing by grinding of supporting rollers to acceptable level
Alarming levels	Increase of 0.1mm/year to do corrective actions in short term Absolute value of 1mm do resurfacing

Table 7.3 Proactive actions against rollers hot bearing

Failure mode	Hot bearing
Root causes	Overloading of rollers (bearing) due to kiln crank Failure of lubrication (* it is not a subject of discussion in the book)
1st Stage-Immediate actions	– Make temperature profile of the journal and thrust disk to determine the source of the temperature increase. – Decrease the load on the supporting station by axial movement of the kiln, – Provide inflow of fresh oil to support the breaking oil film in the bearing – Reduce overall load by decreasing load in the kiln and speed, but do not stop it as oil must circulate inside the bearing – Keep clean the station surrounding (regularly by vacuum) – Eliminate kiln crank (if thermal one)
2nd Stage-actions	Re-adjustment of the roller position to decrease the load on the corresponding journal bearing
Alarming levels	Increase of 5°C/hour to do corrective actions in short term $S_{MIN\text{-}ALARM} < S_{MIN} - ho$ (ho=0,205mm for examined kiln rollers-to be determined based on journal surface finish for every journal-bearing set Absolute value of 46-50°C/60°C (Al/Cu-Sn) for roller adjustment

7.6.3 Proactive Actions for Kiln Crank

Kiln crank is a complex phenomenon already well-elaborated previously. To determine corrective actions based on data, the proactive monitoring as described in this chapter, some additional analysis had to be performed, especially to determine alarm limits for its recognition as we are dealing with online monitoring that gives us possibility not to "wait" till more extensive crank is formed.

7.6.3.1 Setting Alarm Limits for Kiln Crank Condition

Following recommendation from world recognized kiln manufacturing companies as Polysius, FLS, the kiln is said to have crank if the rollers deflection coming from the kiln is more than ± 0.2 mm/0.15 mm. Considering that determination of this value requires some mathematical transformations and analysis, the easiest way to generate an alarm is to set the absolute MIN–MAX value of the roller displacement.

In the following table, values for MAX–MIN of different rollers and situations are given grouped for normal kiln operation and crank condition (Table 7.4) for the exemplary kiln.

Proposal of the authors is to set starting alarm when $S_{\text{peak-to-peak}}$ reaches values of 1,000 μm for 2L. It is because the O_R (Out-of-roundness) of this roller is 700 μm, flexible deflection according to FEM model is up to 300 μm, so any deflection above the two of them (700 + 300 μm) comes from the nontypical operational situations and loads. For 2D (right-hand side) O_R = 170 μm, the alarm limit has

Table 7.4 Parameters describing kiln crank condition

2L	Normal condition			Crank		
No.of measurement	1	2	3	4	5	6
MAX (micrometers)	3759,5	3581,6	3855,3	4294,8	4851,3	4214,5
MIN (micrometers)	3039,4	2853,6	2966,1	3065,0	2978,6	2977,7
MAX-MIN (micrometers)	720,1	728,0	889,2	1229,7	1872,7	1236,7
Kiln magnitude (FFT)	29,0	46,6	54,3	104,4	258,4	76,3
Roller magnitude (FFT)	55,7	50,4	41,2	80,6	51,9	63,0

2D	Normal condition			Crank		
No.of measurement	1	2	3	4	5	6
MAX (micrometers)	2603,9	3021,4	3030,4	3604,7	3855,9	4890,6
MIN (micrometers)	2415,5	2781,5	2766,5	3215,6	3224,0	4558,2
MAX-MIN (micrometers)	188,4	239,9	263,9	389,1	631,9	332,4
Kiln magnitude (FFT)	11,8	46,6	20,0	76,3	87,6	21,6
Roller magnitude (FFT)	14,3	50,4	17,6	63,0	15,4	17,9

Table 7.5 Proactive actions against kiln crank

Failure mode	Kiln crank
Root causes	Overheating of kiln shell causing shell deformation and eccentricity Wrong assembly (**It is not a subject in this book)
1st Stage-Immediate actions	Correct uneven temperature around shell circumference by process actions (change of burner position, increase/decrease sintering zone temperature etc) or forced cooling etc,
2nd Stage- actions	Stop the kiln, shut down coatings with industrial gun or as final thermally re-deform shell
Alarming levels	To be determined for each kiln roller. For the kiln under examination are: 2L: $S_{pk\text{-}pk}$ = 1000 μm; 2D: $S_{pk\text{-}pk}$= 450 μm

been set to 450 μm. For every roller, this value is different as in this process, the roller carrying surface irregularities also participate in the overall measurement. However, after the starting signal, it is recommended to perform short analysis by the relevant engineer and decide on actions. Proactive actions are shown in Table 7.5.

References

1. Z. Stamboliska, Proactive method of low-speed machines condition monitoring. PhD thesis, Wroclaw University of Technology, Poland, 2011
2. R.P. Chapman, *Recommended Procedures for Mechanical Analysis of Rotary Kilns* (Fuller Company, Bethlehem, 1985)
3. FLSmidth Institute, The International Maintenance Seminar, Copenhagen, Vol. I, 2005
4. FLSmidth Institute, The International Maintenance Seminar, vol II (Copenhagen, 2005)
5. Polscan service report, Polysius, Titan cement Usje plant, Kiln no. 3 & 4, 2006
6. E. Rusinski, Z. Stamboliska, P. Moczko, Proactive condition monitoring of kiln roller plain bearings. X International Conference CAE2010 Wroclaw (Poltegor-Instytut, Wroclaw)
7. E. Rusinski, Z. Stamboliska, S. Milevski, Proactive condition monitoring of low-speed machines in cement industry. X International Conference CAE2010 Wroclaw (Poltegor-Instytut, Wroclaw)
8. Siemens: SIREC Manual, Internet Sources
9. Zbigniew Krystowczyk, Geoserverex, Poland: Geometry Measurements of Kiln shell in Dynamic Conditions, Cement & Building Materials No.16-June/2004
10. Polysius: Technology Forum, Neu Beckum, 2004
11. Polscan service report, Polysius, Titan cement Usje plant, Kiln no. 3 & 4, 2010
12. Polscan service report, Polysius, Titan cement Usje plant, Kiln no. 3 & 4, 2008

Index

Z. Stamboliska et al., *Proactive Condition Monitoring of Low-Speed Machines*,
DOI 10.1007/978-3-319-10494-2

MIX
Papier aus verantwortungsvollen Quellen
Paper from responsible sources
FSC® C105338

If you have any concerns about our products,
you can contact us on
ProductSafety@springernature.com

In case Publisher is established outside the EU,
the EU authorized representative is:
Springer Nature Customer Service Center GmbH
Europaplatz 3, 69115 Heidelberg, Germany

Printed by Libri Plureos GmbH
in Hamburg, Germany